AF811038

Haircutting Robots

Steven Li

Haircutting Robots

Springer

Steven Li
Faculty of Information Technology
and Electrical Engineering
University of Oulu
Oulu, Finland

ISBN 978-3-031-84025-8 ISBN 978-3-031-84026-5 (eBook)
https://doi.org/10.1007/978-3-031-84026-5

© The Editor(s) (if applicable) and The Author(s), under exclusive license to Springer Nature Switzerland AG 2025

This work is subject to copyright. All rights are solely and exclusively licensed by the Publisher, whether the whole or part of the material is concerned, specifically the rights of translation, reprinting, reuse of illustrations, recitation, broadcasting, reproduction on microfilms or in any other physical way, and transmission or information storage and retrieval, electronic adaptation, computer software, or by similar or dissimilar methodology now known or hereafter developed.
The use of general descriptive names, registered names, trademarks, service marks, etc. in this publication does not imply, even in the absence of a specific statement, that such names are exempt from the relevant protective laws and regulations and therefore free for general use.
The publisher, the authors and the editors are safe to assume that the advice and information in this book are believed to be true and accurate at the date of publication. Neither the publisher nor the authors or the editors give a warranty, expressed or implied, with respect to the material contained herein or for any errors or omissions that may have been made. The publisher remains neutral with regard to jurisdictional claims in published maps and institutional affiliations.

This Springer imprint is published by the registered company Springer Nature Switzerland AG
The registered company address is: Gewerbestrasse 11, 6330 Cham, Switzerland

If disposing of this product, please recycle the paper.

Preface

The rapid development of robotics is transforming our lives at an unprecedented pace, and haircutting robots stand out as an exciting frontier in this wave of innovation. Combining traditional craftsmanship with cutting-edge technology, haircutting robots embody the pursuit of efficiency, convenience, and personalized services. While no commercially available haircutting robots exist as up to the year 2024, the trajectory of technological advancements suggests their arrival is imminent.

This book aims to fill a gap in the existing literature by providing a systematic exploration of the technologies behind haircutting robots. As the author, I approach this subject from a multidisciplinary perspective, drawing insights from robotics, artificial intelligence, computer vision, and aesthetic design. My goal is to provide readers with a comprehensive understanding of the technical foundations, challenges, and opportunities in this emerging field.

The inspiration for this book stems from the recognition that robotics and artificial intelligence are reshaping industries and professions, from manufacturing to health care. Haircutting robots are a natural extension of this trend, offering solutions to challenges such as the high costs of services in urban centers, exacerbated by rents and living expenses.

This book is designed to equip engineers with both haircutting techniques and the scientific and engineering knowledge needed to inspire the creation of commercial haircutting robots. These skills will help engineers develop haircutting robots that offer practical value. Looking further into the future, advancements in artificial intelligence and robotics will pave the way for fully autonomous haircutting robots. These machines could free barbers from repetitive tasks, allowing them to focus on more creative and fulfilling aspects of their profession.

This book leverages state-of-the-art technologies, including tools like ChatGPT, which has been instrumental in organizing the manuscript. As a testament to the broader impact of artificial intelligence, this book also illustrates how these tools are set to influence various industries, including the emerging domain of haircutting robots.

I hope this book serves as a valuable resource for researchers, entrepreneurs, and professionals interested in the field, providing a detailed breakdown of the core technical components and offering guidance for future development. Together, let us envision a future where haircutting robots become an integral part of our daily lives, blending technology with creativity to meet evolving societal needs.

Oulu, Finland Steven Li

Contents

Overview and Vision of Haircutting Robots

1

1 Introduction to Haircutting Robots

1.1 What is a Haircutting Robot?

A haircutting robot is an intelligent device that integrates knowledge from various disciplines, including robotics, artificial intelligence (AI), computer vision, and aesthetic design. It is designed to provide automated, personalized haircutting services for humans [1, 2]. By sensing, analyzing, and processing the characteristics of a user's head and hair, the robot can execute precise haircuts and styles. It not only possesses high-precision mechanical operation abilities but also features intelligent decision-making and learning capabilities, enabling it to offer customized haircut solutions based on user needs and preferences (Fig. 1).

Haircutting robots have a wide range of applications, covering home, commercial, and specialized environments [3]. In domestic settings, a haircutting robot can offer convenient, safe, and personalized haircut services for family members, particularly suited for households with children, the elderly, or individuals with limited mobility. In commercial environments, robots can provide efficient self-service haircutting in hair salons, shopping malls, airports, and office buildings, reducing customer wait times and improving service efficiency. Additionally, in medical institutions, remote areas, and special environments, haircutting robots can deliver professional haircut services to those with mobility issues or requiring special care, filling gaps in service availability.

1.2 Background of Haircutting Robots

The emergence of haircutting robots is the result of both technological advancements and societal demands. Firstly, the rapid development of robotics technology has laid a solid foundation for the realization of haircutting robots [4]. Advanced

© The Author(s), under exclusive license to Springer Nature Switzerland AG 2025

S. Li, *Haircutting Robots*, https://doi.org/10.1007/978-3-031-84026-5_1

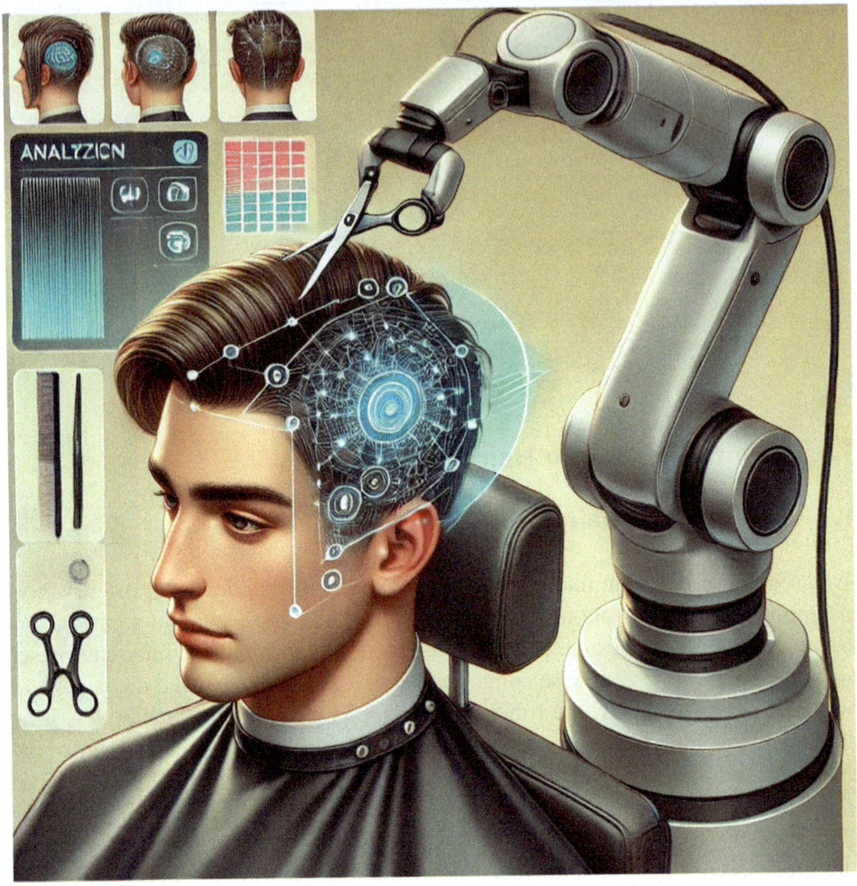

Fig. 1 An illustration of the haircutting robot in operation, sensing the user's head features and performing a haircut

mechanical designs, high-precision control systems, and sensor technologies allow robots to perform more complex and delicate tasks. Mechanical arms are becoming lighter, more precise, and more flexible. The application of servo motors, gear reducers, and advanced control algorithms enables robots to achieve precise motion control (Fig. 2).

Secondly, innovations in AI, particularly breakthroughs in deep learning and reinforcement learning, have equipped machines with the abilities to sense, learn, and make decisions [5, 6]. Significant advancements have been made in image recognition, speech recognition, and natural language processing. The maturity of computer vision technology allows robots to capture and analyze environmental information, performing complex image processing and understanding. These technological advancements provide the foundation for haircutting robots to achieve intelligence.

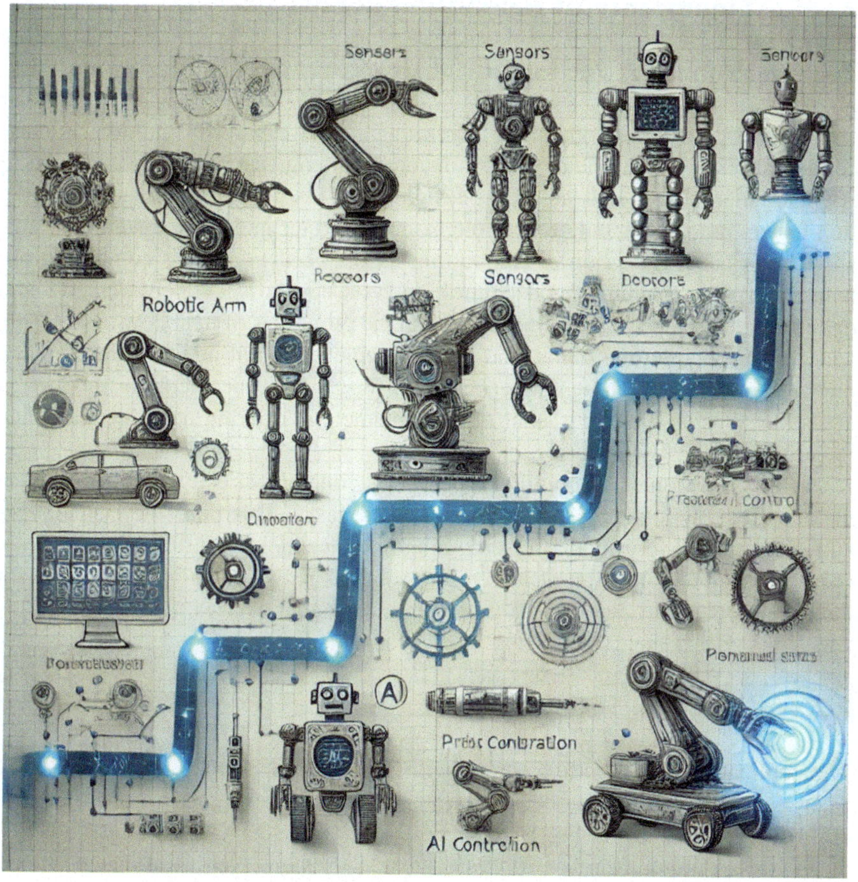

Fig. 2 Branches and development of robotics technology

Moreover, changes in modern lifestyles and the increasing demands of people also create conditions for the development of haircutting robots. As life's pace accelerates, expectations for service efficiency and quality continue to rise, with personalized and customized consumer demands growing [7]. Traditional haircut services may not fully meet these demands in terms of time, location, and service content. The growing elderly population and the increase in people living alone also drive the demand for convenient and reliable self-service options. Haircutting robots can provide efficient, convenient, and personalized haircut services, aligning with the trends of modern societal needs.

2 Significance of Haircutting Robots

2.1 Promoting Technological Development

The research and development of haircutting robots exemplify multidisciplinary collaboration, involving fields such as mechanical engineering, computer science, AI, computer vision, and aesthetic design [8]. This cross-disciplinary cooperation fosters innovation and breakthroughs in related technologies, opening up new application areas for robotics. Haircutting robots need to address complex problems in sensing, planning, and control, such as how to accurately perceive and understand the shape of the human head and the properties of hair, how to plan safe and efficient haircutting paths, and how to achieve natural and friendly human–machine interaction. These challenges push forward advancements in robotics in areas like sensing, planning, control, and interaction, offering new directions and opportunities for researchers.

At the same time, haircutting robots must interact directly with humans in unstructured environments, completing delicate tasks. This poses new challenges and opportunities for robotics. Overcoming these technological difficulties will not only lead to the realization of haircutting robots but also serve as a reference for the development of other service robots and intelligent devices.

2.2 Potential Social Impact

The application of haircutting robots will enhance convenience in people's lives. They can offer haircut services anytime, anywhere, saving users the time and effort of visiting a salon. For busy professionals, the elderly, and individuals with disabilities, haircutting robots provide great convenience [9]. Through personalized services, users can customize their haircutting solutions according to their preferences and needs, improving their quality of life (Fig. 3).

Moreover, the popularization of haircutting robots may lead to changes in professional roles. Barbers' roles may shift from traditional manual operations to the management and maintenance of equipment, as well as providing higher-level aesthetic consultation and hairstyle design services. This transition can help elevate barbers' professional skills and social status, offering new career development paths [10].

The research and application of haircutting robots will also give rise to a series of emerging industries, including intelligent haircutting device manufacturing, customized haircutting service platforms, and the development of related software and applications. The rise of these industries will drive the growth of related fields, creating new job opportunities and stimulating economic growth. Additionally, the widespread use of haircutting robots will promote the standardization and regulation of the industry, contributing to its healthy development.

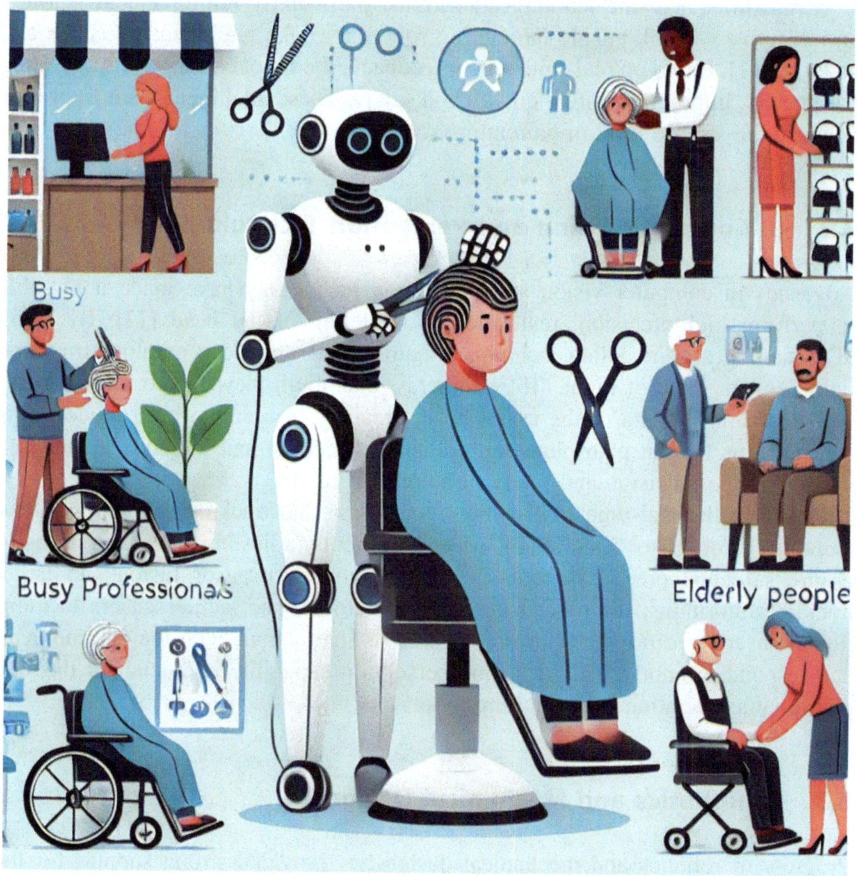

Fig. 3 An illustration showing the convenience haircutting robots provide for different groups of people

3 Latest Technological Advancements in Haircutting Robots

3.1 Artificial Intelligence and Machine Learning Technologies

The progress in AI and machine learning, particularly in deep learning applied to image recognition, enables haircutting robots to accurately capture the user's head and hair information, providing reliable data for haircut planning and execution [5]. Deep learning, especially convolutional neural networks (CNN), has led to significant advancements in image recognition. Haircutting robots can use deep learning to perform facial recognition, head contour detection, and hair feature extraction.

Meanwhile, machine learning algorithms, particularly reinforcement learning and adaptive control, enable haircutting robots to make intelligent decisions and adapt [6, 11]. Based on real-time sensor feedback, the robot can adjust its operating parameters, improving haircut quality and safety. These intelligent features greatly enhance the service level of haircutting robots.

3.2 Computer Vision and Perception Technologies

Advances in computer vision and perception technology have made it possible to perform high-precision, real-time 3D modeling of the head [12]. By using 3D scanning and modeling techniques, haircutting robots can employ structured light sensors, time-of-flight (ToF) cameras, and multi-view stereo vision methods to obtain 3D data of the user's head, creating an accurate head model. This data informs haircut path planning, collision detection, and hairstyle simulation, improving the precision and safety of haircuts (Fig. 4).

Additionally, real-time environment perception and tracking technology allow haircutting robots to detect slight user movements, adjusting their operations to ensure safe and precise haircuts [13]. Through multi-sensor fusion and target tracking algorithms, the robot can use vision, force, and tactile sensors to monitor the user's position and posture, as well as interactions with the environment, and respond promptly. This real-time perception capability is crucial for the safe and reliable operation of haircutting robots.

3.3 Robotics and Mechanical Design

Progress in robotics and mechanical design has provided strong support for the realization of haircutting robots [4]. Advanced robotic arms and end effectors give haircutting robots high degrees of freedom, precision, and flexibility, enabling them to perform complex haircutting motions. The design of robotic arms must consider requirements such as lightweight, high rigidity, and high precision, using advanced materials and structures. End effectors need to simulate the fine motor skills of human hands, securely gripping and manipulating tools such as scissors, clippers, and combs.

Direct-drive motors and precision control technologies have been applied to achieve accurate control of robotic arms and end effectors, improving operational performance and safety [14]. Direct-drive motors are widely used in precision control applications due to their high accuracy, fast response speed, and low noise. In haircutting robots, direct-drive motors are used to drive robotic arms and end effectors, enabling precise control of position, speed, and force. By combining advanced control algorithms such as model predictive control and adaptive control, the performance and safety of the robot can be further improved.

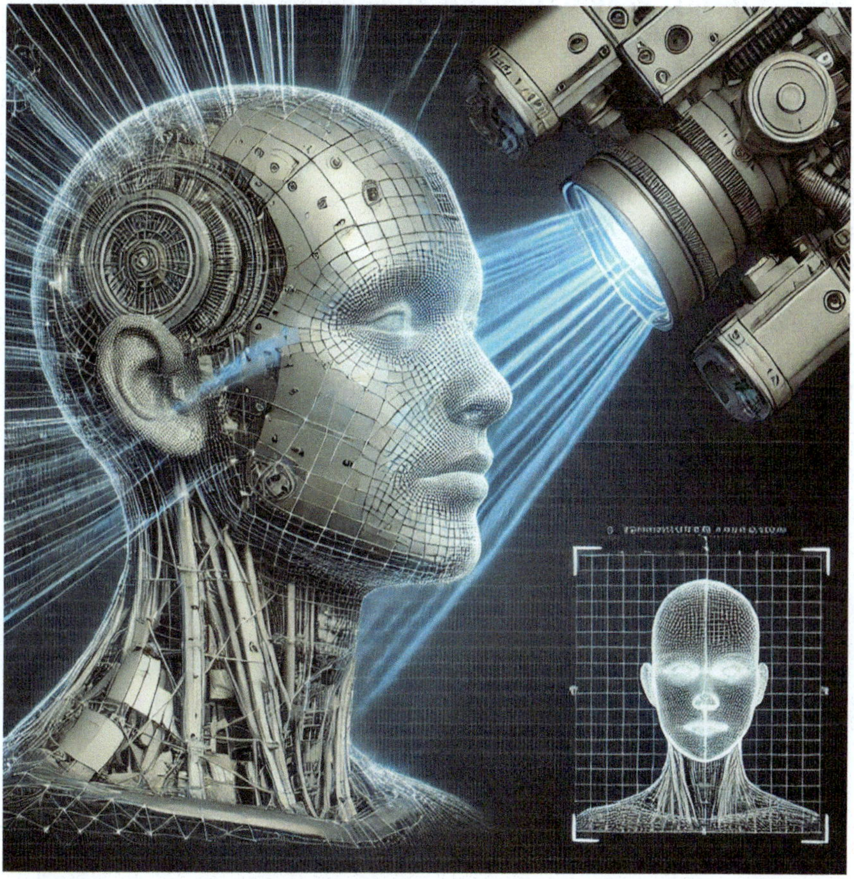

Fig. 4 3D head modeling in haircutting robots

3.4 3D Printing and CNC Machining Technologies

The maturity of 3D printing and CNC machining technologies has made it possible to rapidly manufacture customized haircut tools [15]. Haircutting robots require tools of specific shapes and functions, and through 3D printing, designs can be iterated quickly to meet personalized needs. For example, specially shaped scissors, clippers, and combs can be optimized for robotic operations, improving haircut outcomes.

In addition, rapid manufacturing techniques can accelerate prototype design and testing in the development of haircutting robots. By using a fast design-manufacture-test cycle, research teams can quickly identify issues, optimize designs, reduce development cycles, and improve product quality. This is of great significance for the development of new products and technology iterations.

3.5 Virtual Reality (VR) and Augmented Reality (AR) Technologies

The development of virtual reality and augmented reality technologies offers users immersive hairstyle design and trial experiences [16]. Users can preview different hairstyles on their own head in a virtual environment, adjusting the length, color, and style parameters in real-time. This virtual trial function helps users choose the most satisfying hairstyle before the actual haircut, enhancing the user experience (Fig. 5).

At the same time, AR technology can be used for human–machine interaction during the haircutting process, enhancing the experience. Users can interact with the haircutting robot through gestures or voice commands, adjusting haircut plans or pausing operations. The robot can also display haircut progress, reminders,

Fig. 5 User trying on hairstyles using AR technology

and safety warnings in the user's field of vision. This natural interaction method increases user engagement and control.

3.6 Internet of Things (IoT) and Big Data Technologies

The Internet of Things (IoT) and big data technologies have played an important role in the development of haircutting robots [17]. By connecting devices and enabling remote monitoring, haircutting robots can be networked, allowing operators to monitor device status in real-time, conduct fault diagnosis, and perform maintenance, improving the reliability and maintainability of the equipment. Users can also check the status of the haircutting robot and schedule services through mobile devices.

The collection and analysis of user data allow haircutting robots to provide more personalized services. By analyzing users' haircutting histories, preferences, and feedback through big data technology, haircutting robots can recommend new hairstyles, optimize haircutting processes, and enhance user satisfaction. Additionally, this data provides valuable insights for businesses' market analysis and product improvements (Fig. 6).

4 Vision for Haircutting Robots

4.1 Haircutting Robots in Future Life

In the future, haircutting robots are expected to become essential assistants in people's lives [18]. At home, they can provide convenient, safe, and personalized haircut services for family members, allowing users to enjoy high-quality haircuts at any time based on their schedule. Haircutting robots can also be integrated with other smart home devices, enhancing the level of home automation. For example, they can remind users to get a haircut based on their schedule or play their favorite music during the haircutting process.

In public places such as shopping malls, airports, and office buildings, haircutting robots can offer self-service haircut options. Users can quickly get a haircut while waiting, saving time. This service model not only improves efficiency but also adds a new attraction to commercial spaces. Haircutting robots can also be combined with mobile apps to enable appointment booking, payments, and service reviews, further enhancing the user experience.

4.2 Development Roadmap for Haircutting Robots

The development of haircutting robots requires a clear roadmap. The short-term goal should focus on technology validation and prototype development, breaking through key technologies and optimizing product performance. A fully functional

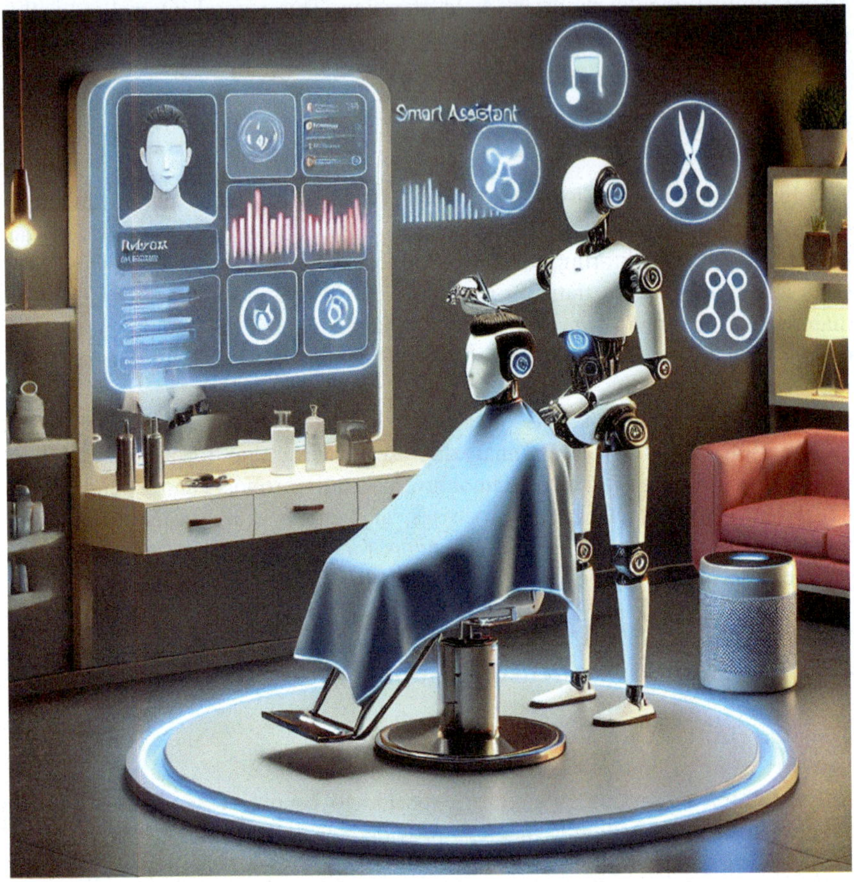

Fig. 6 Application of IoT and big data in haircutting robots

prototype of the haircutting robot should be developed and tested in both laboratory settings and small-scale user trials to collect feedback and improve the design. The objective of this phase is to verify the feasibility of the technology, laying the foundation for commercialization.

In the medium to long term, efforts should focus on productization and commercialization, refining the product design, reducing production costs, and enhancing user experience to bring the product to market. Additional functions, such as beauty treatments, hair dyeing, and hair care, can be incorporated to expand the application scope. For market promotion, a reasonable business model should be developed to explore domestic and international markets, establish brand influence, and build service networks. At the same time, participation in the formulation of industry standards for haircutting robots is necessary to promote the healthy development of the industry.

It is essential to address potential technological challenges, cost control, and legal and ethical issues, and formulate corresponding countermeasures. In terms of technology, further breakthroughs are needed in areas such as perception, planning, control, and human–machine interaction. This can be achieved by strengthening cooperation between industry, academia, and research institutions, attracting experts from various fields to participate in collaborative efforts. For cost control, large-scale production, supply chain optimization, and the application of new materials can help reduce production costs. On the legal and ethical front, actively participating in the development of relevant laws and regulations, and establishing comprehensive user privacy and safety protection mechanisms will enhance user trust.

5 How to Use This Book

This book is suitable for researchers in the field of robotics, professionals in the haircutting industry, researchers in AI and computer vision, product managers, entrepreneurs, and readers interested in haircutting robots. Readers can read through the chapters in order to systematically learn about haircutting robots or choose specific chapters based on their interests and needs for in-depth study, such as mechanical design, control algorithms, human–machine interaction, or market analysis. By combining real-world projects, readers can refer to the methods, case studies, and best practices presented in this book to design and develop haircutting robots. Readers are also encouraged to exchange ideas and practical experiences with peers, experts, and scholars to foster collective progress.

By studying this chapter, readers should gain a comprehensive understanding of the concept, background, significance, and future development of haircutting robots. As an innovative product of multidisciplinary integration, haircutting robots not only contribute to technological progress but also have the potential to bring convenience and innovation to people's lives. In the following chapters, readers will delve deeper into various aspects of haircutting robots.

References

1. Sciavicco, L., & Siciliano, B. (2012). *Modelling and Control of Robot Manipulators*. Springer Science & Business Media.
2. Goodfellow, I., Bengio, Y., & Courville, A. (2016). *Deep Learning*. MIT Press.
3. Yang, G., et al. (2018). Medical robotics—regulatory, ethical, and legal considerations for increasing levels of autonomy. *Science Robotics*, 3(21).
4. Khatib, O. (2014). *Robotics and Autonomous Systems*. In Springer Handbook of Robotics: Springer.
5. LeCun, Y., Bengio, Y., & Hinton, G. (2015). Deep learning. *Nature, 521*(7553), 436–444.
6. Mnih, V., et al. (2015). Human-level control through deep reinforcement learning. *Nature, 518*(7540), 529–533.
7. Pine II, B. J., & Gilmore, J. H. (1999). *The Experience Economy: Work is Theater & Every Business a Stage*. Harvard Business Press.

8. Chen, X., et al. (2020). A survey of robot learning from demonstration. *Robotics and Autonomous Systems*, 122.
9. Broekens, J., Heerink, M., & Rosendal, H. (2009). Assistive social robots in elderly care: A review. *Gerontechnology, 8*(2), 94–103.
10. Frey, C. B., & Osborne, M. A. (2017). The future of employment: How susceptible are jobs to computerisation? *Technological Forecasting and Social Change, 114*, 254–280.
11. Kober, J., Bagnell, J. A., & Peters, J. (2013). Reinforcement learning in robotics: A survey. *The International Journal of Robotics Research, 32*(11), 1238–1274.
12. Zhang, S., & Yau, Y. K. (2010). High-speed three-dimensional shape measurement using structured light methods: A review. *Optics and Lasers in Engineering, 48*(2), 149–158.
13. Brigato, L., & Iocchi, L. (2021, January). A close look at deep learning with small data. In 2020 25th *International Conference on Pattern Recognition (ICPR)*, (pp. 2490–2497). IEEE.
14. Schölkopf, B., & Smola, A. J. (2002). *Learning with Kernels: Support Vector Machines, Regularization, Optimization, and Beyond*. MIT Press.
15. Gibson, I., Rosen, D. W., & Stucker, B. (2014). *Additive Manufacturing Technologies: 3D Printing, Rapid Prototyping, and Direct Digital Manufacturing*. Springer.
16. Azuma, R. T. (1997). A survey of augmented reality. *Presence Teleoperators Virtual Environments, 6*(4), 355–385.
17. Atzori, L., Iera, A., & Morabito, G. (2010). The internet of things: A survey. *Computer Networks, 54*(15), 2787–2805.
18. Wang, T. M., Tao, Y., & Liu, H. (2018). Current researches and future development trend of intelligent robot: A review. *International Journal of Automation and Computing, 15*(5), 525–546.

A Review of Attempts Towards Haircutting Robots

2

The pursuit of robotic haircutting solutions has seen diverse and innovative approaches, ranging from the use of robotic arms to experimental drone-based systems. These attempts reflect a growing interest in leveraging robotics to enhance personal grooming, offering convenience, precision, and efficiency. The field spans a spectrum of methodologies: from collaborative robots capable of sensing and adapting to head contours, to teleoperated systems that combine human expertise with robotic precision, and even conceptual ideas such as haircutting drones. Each approach highlights unique challenges and opportunities, from ensuring safety and accuracy to achieving adaptability and affordability. This chapter provides a comprehensive review of existing attempts to develop haircutting robots, offering insights into current capabilities and the future potential of this evolving field.

1 Helmet for Haircutting

One notable example of imaginative thinking in automated haircutting devices is the "Shaving Helmet," (Fig. 1) created by Think modo as part of a viral marketing campaign for Head Blade razors [1, 2]. Although ultimately revealed to be a hoax, the concept sparked widespread intrigue and discussion, shedding light on the potential for wearable grooming devices. The helmet was envisioned as a sleek, dome-shaped apparatus that could be placed over the head, providing a quick, automated shaving experience in as little as 20 s. Its design incorporated multiple razors positioned strategically within the inner surface of the helmet. These razors were imagined to be suspended on adjustable tracks, allowing them to conform to the natural contours of the scalp for an even and precise shave. The vision extended to the inclusion of advanced components, such as a shaving cream distribution

© The Author(s), under exclusive license to Springer Nature Switzerland AG 2025
S. Li, *Haircutting Robots*, https://doi.org/10.1007/978-3-031-84026-5_2

system, which would dispense an optimal amount of lubricant to ensure smooth operation and minimize irritation.

The razors were conceptualized to operate through motorized mechanisms that mimicked the precision and fluidity of a professional barber's hand movements. Each razor would move independently, guided by sensors capable of detecting the unique shape and size of the user's head. This level of adaptability, while fictional in the context of the "Shaving Helmet," aligns closely with the challenges faced in designing real robotic systems for personal grooming. The imagined system also emphasized speed and convenience, allowing users to achieve a perfectly shaved head without the need for manual effort or traditional grooming tools.

Beyond its fictional practicality, the "Shaving Helmet" serves as a thought-provoking model for what a haircutting robot could potentially achieve. Modern technologies such as LiDAR and 3D vision systems could be integrated into such a device, enabling precise mapping of the scalp's topography to guide razors accurately. Safety mechanisms, such as pressure sensors to avoid cuts and skin irritation, could address the critical issue of user comfort and safety. Furthermore, haptic feedback and real-time adjustments could provide a level of finesse similar to human touch, making the system more reliable and appealing.

The broader impact of the "Shaving Helmet" lies in its ability to inspire a new generation of wearable grooming robots. The concept highlights the potential for blending creativity with robotics, particularly in automating routine tasks like haircuts or shaving. For example, a future "haircutting helmet" could include modular attachments for trimming, layering, or even styling hair, expanding its functionality

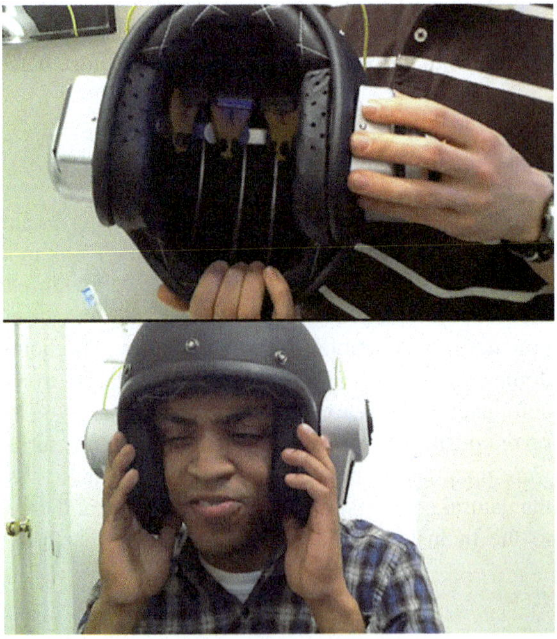

Fig. 1 The imaginative "Shaving Helmet" for automated haircutting created by Thinkmodo as part of a viral marketing campaign for HeadBlade razors, which is ultimately revealed to be a hoax

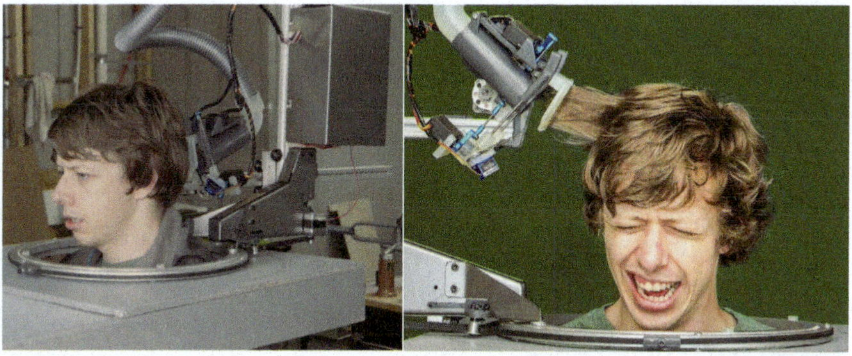

Fig. 2 The first version of the haircutting robot invented by Shane Wighton

beyond simple shaving. Despite its fictional origin, the "Shaving Helmet" captures the public's imagination and demonstrates how bold, unconventional ideas can spark innovation. By pushing the boundaries of what is possible, such concepts challenge engineers and designers to think creatively about the intersection of robotics, personal care, and everyday convenience (Fig. 2).

2 Robot Arms for Haircutting

2.1 Automated Haircutting

Shane Wighton's Haircutting Robots

Shane Wighton, a highly innovative DIY enthusiast and engineer, has developed two experimental versions of a haircutting robot, demonstrating significant progress in the field of automated grooming. The first version [3], showcased in his video titled *"I Built a Robot to Cut My Hair,"* employed a straightforward design that highlighted the foundational concept of a DIY automated haircutting system. This version consisted of a robotic arm equipped with a motorized cutting tool, programmed to execute a series of pre-defined movements to uniformly trim the user's hair. The design required meticulous manual setup, including fixing the cutting path in advance, which made it less adaptable to variations in head shape or hair density. Despite its limitations, the first prototype served as a proof-of-concept, offering insight into how an automated device could replicate basic haircutting tasks. However, it lacked advanced features like real-time adaptability or feedback mechanisms.

In his quest to improve upon the initial concept, Wighton developed a second version [4] (Fig. 3), featured in his video *"I Upgraded My Haircut Robot."* This iteration addressed many of the shortcomings of the first by incorporating cutting-edge technology such as computer vision and real-time head tracking. The system used a camera and software to scan the head's shape, enabling the robot to dynamically

Fig. 3 The second version of the haircutting robot invented by Shane Wighton

adjust its cutting path based on the contours of the user's scalp. This eliminated the need for extensive manual calibration and significantly improved precision. The robotic arm in this version was more sophisticated, featuring smoother movements and greater control over cutting depth and angle, ensuring a more uniform and professional finish. Additionally, the upgraded system was designed to handle variations in hair length and thickness, making it far more versatile than its predecessor.

Both versions of Wighton's haircutting robot serve as a testament to the potential of DIY engineering in solving practical, everyday problems. The first version demonstrated that automated haircutting is feasible even with minimal resources, while the second version pushed the boundaries by integrating advanced robotics and sensing technologies. These projects not only highlight the progress in automated grooming but also inspire further exploration into the possibilities of personalized and efficient robotic solutions for home use. As Wighton continues to refine his designs, his work underscores the transformative potential of combining creativity, engineering, and robotics in addressing routine tasks.

3D Scanning Based Automated Haircutting

An advanced approach to robotic haircutting involves the use of a fully autonomous robotic arm paired with 3D scanning technology to map the head's shape. In the example demonstrated on YouTube [5] (Fig. 4), the system integrates a robotic arm equipped with hair clippers and a smartphone-based 3D scanner. The process begins by using the smartphone to capture a detailed 3D model of the user's head. This model provides the robotic system with precise information about the scalp's contours, allowing it to plan a customized haircut path.

Once the scanning is complete, the robotic arm processes the data to generate a cutting trajectory tailored to the individual's head shape and desired haircut style. The robotic arm, equipped with a cutting tool as its end effector, follows this

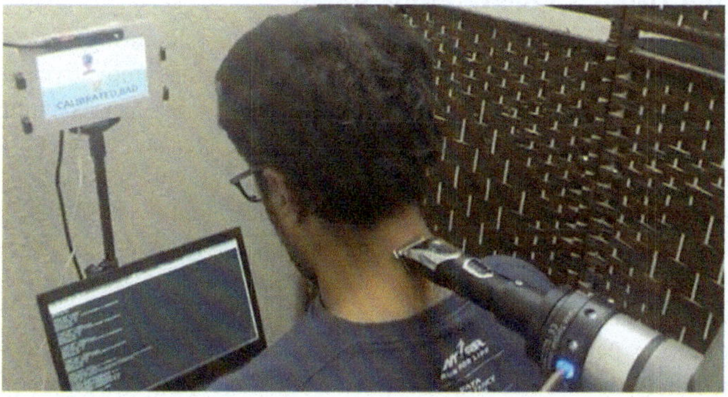

Fig. 4 A type of automated haircutting robot with 3d scanning using a camera on phones

predefined path with high precision. Safety features, such as collision detection and controlled cutting pressure, are integrated to ensure that the process remains safe and avoids damaging the skin.

This system represents a significant leap in autonomous haircutting technology, as it eliminates the need for manual intervention or preprogramed motions. By incorporating 3D scanning, the robot adapts to unique head shapes and personal styling preferences, offering a truly customized experience. The integration of a smartphone for scanning also demonstrates how consumer-grade technology can be effectively used to enhance the capabilities of robotic systems.

This innovative setup not only highlights the potential of combining robotics and 3D imaging but also paves the way for future development of fully autonomous grooming devices. With further refinement, such systems could offer reliable, salon-quality results at home, making them a valuable addition to personal care technologies (Fig. 5).

Fig. 5 Haircutting using a collaborative robot

Haircutting Using a Collaborative Robot

As shown in Fig. 5 [6], the robotic haircutting system featuring the Sawyer collaborative robot from Rethink Robotics demonstrates an innovative approach to automated personal grooming. Designed as a co-bot, Sawyer is equipped with advanced force-sensing capabilities, allowing it to detect and respond to physical contact in real-time. This feature ensures that the robot can "feel" when it touches the user's head, enabling safe and gentle operation. In this case, Sawyer is outfitted with an automatic trimmer as its end effector, precisely manoeuvring around the head to deliver an even haircut. The robot leverages its 7 degrees of freedom and an integrated Cognex vision system to map the contours of the user's scalp and adjust its cutting path dynamically. These features, combined with its ability to stop immediately upon detecting unexpected resistance, make Sawyer particularly suitable for tasks requiring precision and safety. This application highlights the versatility of collaborative robots, bridging the gap between human-like dexterity and automated efficiency, and offers a glimpse into the future of robotic-assisted personal care solutions.

2.2 Teleoperated Haircutting

NBC News Reported Teleoperated Haircutting

Another innovative approach to automated haircutting involves the use of a pre-existing robotic arm, as demonstrated in the case highlighted by NBC News [7] (Fig. 6). In this project, a skilled barber remotely controlled a robotic arm to perform a live head-shaving session. The setup utilized a high-precision robotic arm, commonly found in industrial applications, adapted for personal grooming purposes. Equipped with a clipper as the end effector, the robotic arm was capable of replicating the barber's motions, allowing for real-time hair cutting with remarkable accuracy.

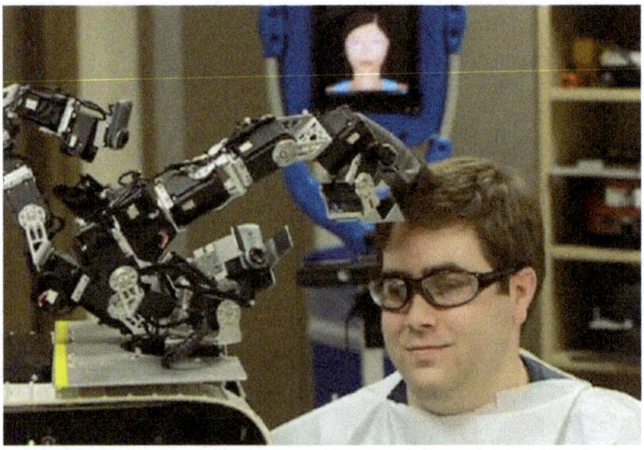

Fig. 6 A type of teleoperated haircutting as reported by NBC News

The process relied on the operator's expertise to guide the robot via remote control, combining the precision of the robotic arm with the barber's skills. Cameras and sensors enabled the barber to monitor the head's contours and adjust the arm's movements dynamically to achieve a clean and even shave. This innovative demonstration not only showcased the versatility of robotic technology in unconventional applications but also illustrated how remote-controlled systems could be used to extend human expertise to tasks requiring precision and adaptability.

The project was executed as a charity initiative, drawing significant attention for its creative application of robotics in everyday tasks. While not fully autonomous, this method demonstrated the feasibility of integrating human skill with robotic precision, offering potential insights into future advancements in personal grooming technology. Such an approach could pave the way for hybrid systems where professionals remotely operate robotic devices, delivering high-quality results without requiring their physical presence.

Teleoperated Robotic Haircutting with High Precision

In the video [8], a person demonstrates an impressive teleoperated robotic system for haircutting. Using a robotic manipulator equipped with scissors, the individual operates the robot to cut their own hair from behind. The teleoperation setup allows the user to control the robot's movements with precision, showcasing smooth and deliberate motion throughout the demonstration. The robotic arm is shown carefully navigating the contours of the operator's head, trimming sections of hair with remarkable accuracy. This high level of control is made possible by the responsive nature of the robotic arm and the operator's skill in telemanipulation (Fig. 7).

The demonstration highlights how teleoperation bridges the gap between human expertise and robotic dexterity. The operator maintains complete command over the robot, allowing for subtle adjustments in real-time to ensure the haircut is symmetrical and consistent. Despite the inherent challenges of cutting hair without a

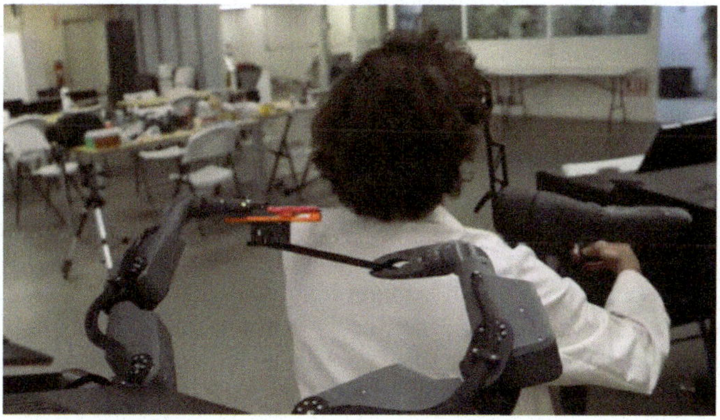

Fig. 7 Teleoperated robotic haircutting with high precision

direct line of sight, the system's design and control interface enable the operator to achieve a professional-quality result. This suggests the robot's potential for broader applications in personal grooming tasks.

What stands out in this demonstration is the smoothness and reliability of the robotic arm's movements. The system effectively translates the operator's intent into precise, controlled actions, minimizing errors and ensuring safety. By leveraging teleoperation, this approach avoids the complexity of fully autonomous robotic haircutting while still providing a high level of precision and adaptability. Such a system could be particularly beneficial for scenarios where professional assistance is unavailable, or for those seeking a novel, efficient way to manage personal grooming tasks. The successful execution of this demonstration underscores the promise of teleoperated robotics in achieving tasks that demand both fine motor skills and human oversight.

2.3 Panasonic's Shampoo and Head Massage Robot

Although not a haircutting robot, Panasonic's shampoo and head massage robot [9, 10] provides valuable inspiration and shares functional similarities with automated grooming devices. Demonstrated in two videos, this robot is designed to wash and massage the user's scalp, combining mechanical efficiency with human-like precision. The system uses a series of articulated robotic arms equipped with 16 finger-like extensions that simulate the motions of a professional scalp massage. These robotic "fingers" are capable of adjusting pressure and motion dynamically, ensuring a gentle yet thorough cleaning and relaxing experience.

The robot's operation begins with a scanning phase, where it maps the user's head to understand its shape and contours. This allows the system to adapt its movements to the individual's scalp, ensuring optimal comfort and effectiveness. Once the scanning is complete, the robotic arms work in tandem to deliver a synchronized washing and massaging sequence. This process mimics the actions of human hands, using a combination of circular motions, kneading, and gentle scrubbing (Fig. 8).

The design philosophy behind this robot aligns closely with the goals of a haircutting robot: providing a precise, adaptable, and safe experience tailored to each user. Features like head shape mapping, adjustable pressure, and multi-point coordination could be directly applied to automated haircutting devices. For example, a similar scanning process could be used in a haircutting robot to ensure accurate hair length adjustments and safe navigation of cutting tools. Additionally, the emphasis on user comfort and relaxation highlights the importance of balancing efficiency with a pleasant experience in personal care robotics.

Panasonic's shampoo robot serves as a benchmark for how robotics can replicate intricate human tasks, making it a compelling example for designers of

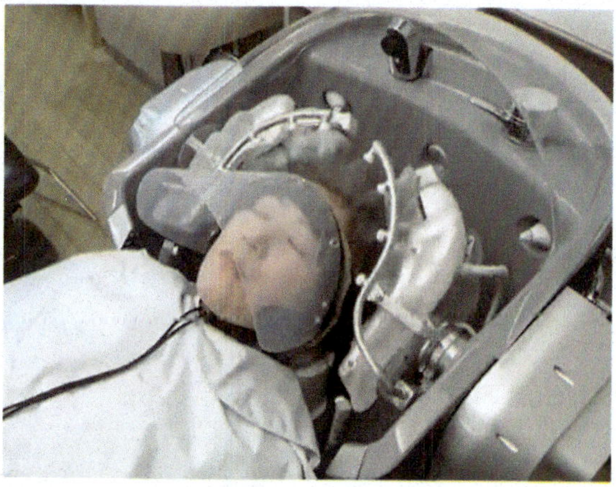

Fig. 8 Panasonic's shampoo and head massage robot

automated grooming technologies. By incorporating similar adaptive and user-focused features, haircutting robots could offer not only functional excellence but also a comfortable and personalized service, transforming the future of personal grooming.

3 Drones for Haircutting

The idea of using drones for haircutting, while still in its infancy, offers a fascinating glimpse into how advancements in robotics and control systems could transform grooming technologies. A video [11] showcasing an experimental attempt at drone-based haircutting demonstrates the current state of this concept. In the demonstration, a drone is equipped with cutting tools and maneuvered to trim hair. While the results are rudimentary and far from practical application, the experiment serves as a proof of concept for how aerial robots could potentially be adapted for such tasks.

The primary challenge in drone-based haircutting lies in achieving the necessary precision. Current drone control systems lack the millimeter-level accuracy required for safe and precise haircutting, but with ongoing improvements in drone stabilization, sensor integration, and control algorithms, such precision could become achievable. For instance, future drones equipped with advanced LiDAR or vision systems could scan and model the user's head in real-time, enabling them to adjust their movements dynamically. Additionally, AI algorithms could provide real-time path planning to ensure even and safe trimming, similar to how autonomous robotic arms currently function (Fig. 9).

Using drones instead of robotic arms for haircutting offers several potential advantages. Drones are inherently more flexible and mobile, capable of accessing angles and positions that might be challenging for fixed or grounded systems. This mobility could allow for faster operation and reduced setup complexity. Moreover,

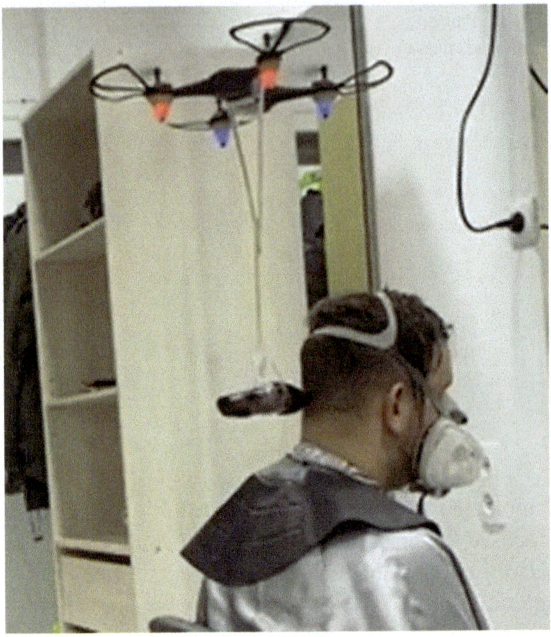

Fig. 9 Using drones for haircutting

drones could provide a cost-effective alternative to mechanical arms, especially as their precision improves and their size and energy consumption decrease.

While the technology is far from maturity, the concept of drone-based haircutting holds significant promise. With advances in control accuracy, safety mechanisms, and autonomous functionality, drones could one day offer a highly flexible and efficient solution for automated haircuts. This potential makes it an exciting avenue for future research and development in personal grooming robotics, combining aerial robotics with the precision of modern automation technologies.

4 Conclusions

As we have seen, the journey toward a fully functional haircutting robot is marked by creativity, experimentation, and technological innovation. From tele operated robotic arms delivering precise trims to conceptual drone systems with the promise of flexibility and cost-effectiveness, these attempts underscore the transformative potential of robotics in personal grooming. While many of these approaches are still in their infancy, they provide a glimpse into a future where haircutting robots could become a common reality, combining safety, precision, and personalization. By understanding and learning from these pioneering efforts, we pave the way for more refined and practical solutions, pushing the boundaries of what robotics can achieve in everyday life.

References

1. https://www.youtube.com/watch?v=5bgRszdUdhQ
2. https://www.youtube.com/watch?v=3ZRAMy0SizY
3. https://www.youtube.com/watch?v=7zBrbdU_y0s&t=578s
4. https://www.youtube.com/watch?v=WQ8Xgp8ALFo&t=583s
5. https://www.youtube.com/watch?v=jY_gpi_gsRI
6. https://www.youtube.com/watch?v=-cgU1RgjWXw
7. https://www.nbcnews.com/tech/tech-news/robot-barber-shaves-human-head-charity-flna510080
8. https://www.youtube.com/watch?v=RXhBjsDLKDM
9. https://www.youtube.com/watch?v=aWxF_ZXs-_o
10. https://www.youtube.com/watch?v=5EbTOQ6uD1k
11. https://www.youtube.com/shorts/9RHroJPhlKw

Fundamentals of Haircutting Techniques

3

The core of haircutting techniques lies in understanding the structure of hair growth, the relationship between facial features and hairstyle design, and the tools and technical support required during the haircutting process. By mastering these foundational concepts, barbers can provide precise and personalized haircut services based on each client's unique needs.

1 Basic Principles and Process of Haircutting

1.1 Hair Growth and Structure

Hair is a fibrous structure composed of keratin, consisting mainly of the hair shaft and hair follicle. The hair follicle is the foundation of hair growth, with a complex structure that includes the hair papilla, hair matrix, and sebaceous glands [1]. The health of the hair follicle directly determines the quality and speed of hair growth. The hair growth cycle generally consists of three phases: the anagen (growth phase), catagen (regression phase), and telogen (resting phase), each lasting between two and six years [2]. This cyclical change means that hair is not constantly growing but undergoes a natural process of shedding and regeneration [3].

Different groups of people have varying hair types, primarily influenced by genetics and ethnicity. The differences between straight, wavy, and curly hair are fundamentally due to the shape of the hair follicle and the cross-section of the hair [4]. These characteristics determine the texture of each person's hair and the ease of styling. Barbers must understand these differences to make the best judgment when designing hairstyles based on the client's hair texture and structure (Fig. 1).

© The Author(s), under exclusive license to Springer Nature Switzerland AG 2025
S. Li, *Haircutting Robots*, https://doi.org/10.1007/978-3-031-84026-5_3

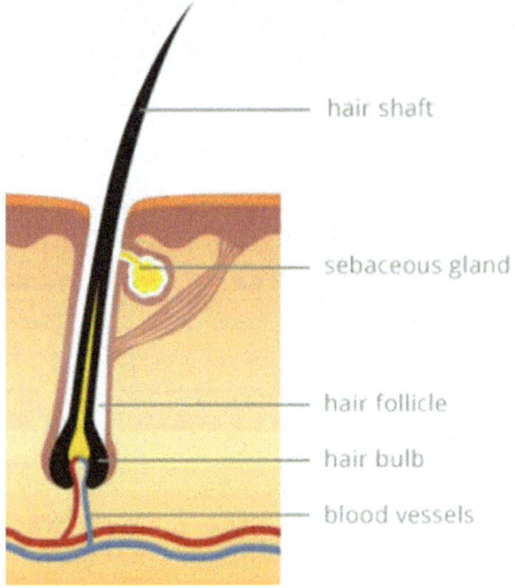

Fig. 1 The structure of hair

1.2 Facial Features and Hairstyle Design

Everyone's facial features are unique, so hairstyle design goes beyond cutting hair. It requires reasonable adjustments based on the individual's face shape, head shape, and the characteristics of their facial features to achieve the effect of enhancing facial proportions and highlighting personal characteristics [5]. For example, for a round face, the hairstyle should focus on elongating facial proportions, while for a square face, soft layers can be used to soften facial lines [6]. An oval face, being more balanced in proportion, can accommodate a variety of hairstyles, whereas a long face benefits from adding horizontal lines to visually shorten the face [7]. For heart-shaped faces, the hairstyle should add volume around the chin to balance the width of the upper and lower parts of the face.

During the hairstyle design process, barbers must consider not only the client's facial contours but also their lifestyle, personality, and profession to provide professional advice. This personalized design process is one of the most challenging and creative aspects of haircutting services (Fig. 2).

1.3 Basic Process of Haircutting

Although the entire haircutting process may seem simple, each step requires precise operation and thorough communication with the client. First, the barber must understand the client's needs, lifestyle habits, and haircutting preferences through conversation [8]. This step is not just about cutting hair but also about providing personalized service. Next comes the design and communication phase, where the barber typically offers professional hairstyle suggestions based on the client's face

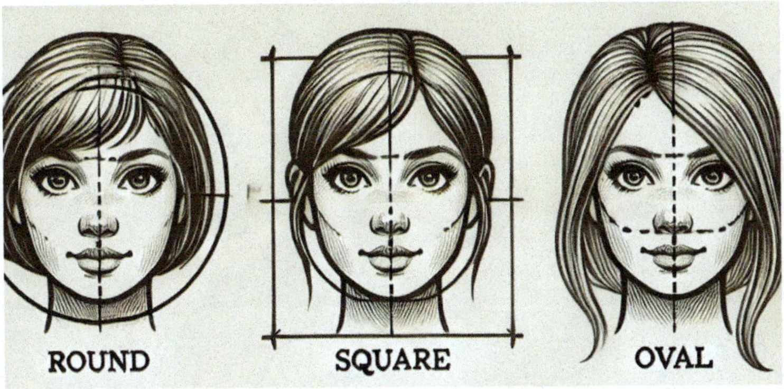

Fig. 2 Suitable hairstyles for different face shapes

shape, hair characteristics, and needs, possibly using auxiliary software or images to display the hairstyle options [9].

The actual haircutting process begins with washing the hair, which not only cleans the hair but also prepares it for the upcoming cut. The barber will divide the hair into sections for easier, gradual cutting [10]. The handling of each section may differ depending on the client's hairstyle requirements and hair texture. During the cutting process, various techniques, such as horizontal cutting, vertical cutting, and diagonal cutting, are used to ensure the overall layering and styling effect of the haircut [11].

The final step in the haircutting process is finishing and styling. The barber will use tools such as blow dryers and styling products based on the client's preferences to set the hairstyle and make detailed adjustments to perfect the overall effect [12]. Finally, the barber will let the client check the results, ask for feedback, and make any necessary adjustments. The client's feedback during this process is crucial to the final result of the haircut (Fig. 3 and Tables 1 and 2).

2 Common Haircutting Tools and Equipment

During the haircutting process, tools are essential to completing each stage. Scissors are the most common tool and are primarily used for basic cutting operations [13]. Straight scissors are mainly used for creating clean lines, while thinning scissors are used to reduce hair volume and create layers [14]. For certain specific hairstyles, curved scissors are helpful for trimming bangs and creating curved lines [15] (Fig. 4).

Electric clippers are often used for men's hairstyles or situations that require quick trimming [16]. Modern clippers not only allow for length adjustment but also provide precision cutting by using interchangeable guard combs [17]. Combs are also indispensable tools in haircutting, helping barbers comb through hair and offering support for sectioning and controlling hair volume during cuts [18].

Fig. 3 The haircutting process

Table 1 Hair types and characteristics

Hair type	Description	Suitable haircutting techniques
Straight	Smooth, shiny, low elasticity	Add layers to increase movement
Curly	High curl, strong elasticity	Thin out to control volume
Wavy	Between straight and curly hair	Emphasize texture, enhance depth

Table 2 Common cutting techniques and application scenarios

Technique	Method	Application scenarios
Point cut	Cut hair ends with the tips of scissors	Adds lightness, softens hair ends
Slide cut	Use scissors to slide through the hair	Creates natural layers, reduces bulk
Thinning	Use thinning scissors to reduce volume	Controls volume, balances head shape

Fig. 4 Types of common scissors

In addition to scissors and clippers, blow dryers and styling tools play an equally important role. Blow dryers are not only used for drying hair but are also critical for setting styles [19]. Curling irons and flat irons are essential tools for barbers when creating personalized hairstyles for clients [20]. Additionally, auxiliary tools such as hair clips, spray bottles, and capes ensure a smooth and hygienic haircutting process [21].

3 Essential Skills and Techniques for Barbers

To become a skilled barber, one must not only have solid tool-handling skills but also master various cutting and styling techniques [22]. The basic techniques of holding scissors and combs are fundamental to the cutting process. Barbers must undergo years of training to ensure precision and flexibility in hand movements [23]. Additionally, the coordination between scissors and combs determines the layering and detail accuracy of the hairstyle [24] (Fig. 5).

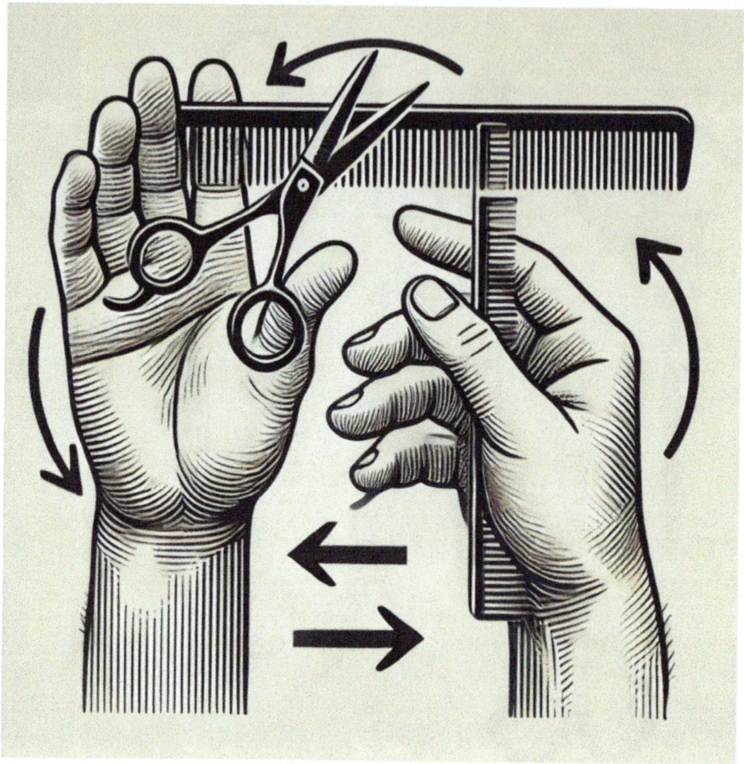

Fig. 5 Correct holding positions for scissors and combs

During the actual haircutting process, barbers must apply different techniques based on the specific hairstyle requirements [25]. Horizontal cuts are suitable for adding weight to the hairstyle, while vertical cuts can elongate the face and create natural layers [26]. Techniques such as slide cutting and point cutting are used to soften hair ends, reduce bulk, and create a lighter, more natural look [27]. By combining these techniques, barbers can craft hairstyles that align with the client's personal style and current fashion trends [28].

Creative hairstyle design depends on understanding the client's personality and current trends. Barbers need to stay updated with the latest fashion trends and learn to strike a balance between tradition and modernity, innovatively designing hairstyles [29]. Moreover, color matching is another critical aspect of hairstyle design. Barbers need to select appropriate hair colors based on the client's skin tone and temperament, using techniques such as gradients and highlights to add flair to the hairstyle [30, 31].

Communication with clients is another essential skill during the haircutting process [32]. By patiently listening to the client's needs and providing professional suggestions based on expertise, barbers can ensure client satisfaction while enhancing their professional image [33].

Formula: The Golden Ratio in Hairstyle Design

In hairstyle design, the golden ratio (1:1.618) is often used to adjust the proportion between the face and the hairstyle, achieving aesthetic harmony [34]. This formula expresses the relationship between hairstyle length and face length as follows:

$$\frac{\text{Hairstyle Length}}{\text{Face Length}} = 1.618$$

By applying this formula, barbers can adjust the hairstyle length to meet aesthetic standards, creating a sense of visual harmony for the client.

References

1. Spong, M. W., Hutchinson, S., & Vidyasagar, M. (2020). *Robot modeling and control.* Wiley.
2. Stenn, K. S., & Paus, R. (2001). Controls of hair follicle cycling. *Physiological Reviews, 81*(1), 449–494.
3. Loussouarn, G., El Rawadi, C., & Genain, G. (2005). Diversity of hair growth profiles. *International Journal of Dermatology, 44*, 6–9.
4. Mellmann, K. (2013). On the emergence of aesthetic illusion an evolutionary perspective. In *Immersion and distance* (pp. 65–88). Brill.
5. Frangie, C. M., Botero, A. R., Hennessey, C., Lees, M., Sanford, B., Shipman, F., & Wurdinger, V. (2007). *Milady standard cosmetology.* Thomson Delmar Learning.
6. Seo, K. J., Lee, H. W., & Kim, M. Y. (2021). Analysis of the psychological effects of beauty therapy using beauty management programs. *The Research Journal of the Costume Culture, 29*(5), 719–734.
7. Williams, T. (2013). Haircutting: Level 2: The interactive textbook. Cengage Learning.
8. Haircutting, Certified Learning in Cosmetology, Hardcover:244 ISBN-13:978-0972151108.
9. Sari, D. M., & Fauzani, A. R. (2023). Analysis of basic hair cutting skills in cosmetic education students. *Edukasi, 17*(2), 212–218.
10. Arden Magtiza, A., & Franceschini, G. (2021). *Barbering techniques for hairstylists.* BCcampus.
11. Spear, J. (2009). *Haircutting for dummies.* World of Books Ltd.
12. Ward, K., Bertails, F., Kim, T. Y., Marschner, S. R., Cani, M. P., & Lin, M. C. (2007). A survey on hair modeling: Styling, simulation, and rendering. *IEEE Transactions on Visualization and Computer Graphics, 13*(2), 213–234.
13. Knight, R. (2015). *Haircutting basics: An easy, step-by-step guide to cutting hair the professional way.* Beauty Craft.
14. Elliott, B. (2007). *Textured tresses: the ultimate guide to maintaining and styling natural hair.* Black Beauty Books.
15. Anderson, S. (2010). *Creative hair styling.* Gingertown Books. ISBN-13: 979-8323157259.
16. Gokdag, R. (2016). The comparison of hairdresser-customer communication: The analysis of communication themes in barbershops. *International Journal on New Trends in Education & Their Implications, 7*(4), 13–20.
17. Pavey, J. (2012). *Milady Standard Cosmetology.* Cengage Learning.

18. Palladino, L., & Green, M. (2015). *Hairdressing and barbering, the foundations: The official guide to level 2, cengage learning EMEA.*
19. Broekens, J., Heerink, M., & Rosendal, H. (2009). Assistive social robots in elderly care: A review. *Gerontechnology, 8*(2), 94–103.
20. Frey, C. B., & Osborne, M. A. (2017). The future of employment: How susceptible are jobs to computerisation? *Technological Forecasting and Social Change, 114*, 254–280.
21. Livio, M. (2002). *The golden ratio: The story of phi, the world's most astonishing number.* Broadway Books.
22. Kuber, J., Bagnell, J. A., & Peters, J. (2013). Reinforcement learning in robotics: A survey. *International Journal of Robotics Research, 32*(11), 1238–1274.
23. LeCun, Y., Bengio, Y., & Hinton, G. (2015). Deep learning. *Nature, 521*(7553), 436–444.
24. Mnih, V., et al. (2015). Human-level control through deep reinforcement learning. *Nature, 518*(7540), 529–533.
25. Swami, V., & Furnham, A. (2008). *The psychology of physical attraction.* Routledge.
26. Cagle, A. (2016). *Fundamentals of hair cutting.* Mindbridge Press.
27. *Haircutting a philosophy and science to precision and texturized haircutting.* (2012). CreateSpace Independent Publishing Platform. ISBN-13: 978-1466263154.
28. Zhang, S., & Yau, Y. K. (2010). High-speed three-dimensional shape measurement using structured light methods: A review. *Optics and Lasers in Engineering, 48*(2), 149–158.
29. Delgado, J. M. D., & Oyedele, L. (2021). Deep learning with small datasets: Using autoencoders to address limited datasets in construction management. *Applied Soft Computing, 112*, 107836.
30. Atzori, L., Iera, A., & Morabito, G. (2010). The internet of things: A survey. *Computer Networks, 54*(15), 2787–2805.
31. Moravec, H. (2009). Rise of the robots-the future of artificial intelligence. *Scientific American, 23.*
32. Goodfellow, I., Bengio, Y., & Courville, A. (2016). *Deep learning.* MIT Press.
33. Lynch, K. M. (2017). *Modern robotics.* Cambridge University Press.
34. Khatib, O. (2014). *Robotics and autonomous systems.* Springer.

Hairstyle Design and Classification 4

1 Characteristics of Various Hairstyles and Suitable Groups

Short hair has always been the top choice for many individuals seeking a clean and sharp style. This type of hairstyle is not only easy to maintain but also effectively enhances facial contours. Common short hairstyles include crew cuts, bob cuts, and layered short hair. The crew cut is typically suitable for expressing a gender-neutral style for both men and women, presenting a clean and crisp look. The bob cut is particularly popular among modern women, showcasing an independent and fashionable aura. Layered short hair creates a sense of lightness through different layer cuts and is suitable for women with thicker hair, making the hairstyle appear more dynamic and three-dimensional (Fig. 1).

The modifying effect of short hair lies in its ability to balance facial proportions. For individuals with round or square faces, short hair can soften overly gentle or prominent facial lines, creating a more harmonious visual effect. Professional women and athletes often choose this hairstyle for the convenience it offers in their work and lives. Research shows that short hair not only conveys a competent image but also enhances self-confidence and boosts work efficiency [1].

Medium-length hair, positioned between short and long hair, retains the elegance of long hair while offering the manageability of short hair. Common forms include straight hair, wavy hair, and layered hair. Straight hair usually appears simple and elegant, well-suited for various occasions and styles. Wavy hair adds a sense of movement through its natural curls, presenting a casual yet elegant vibe. Layered hair enhances lightness by trimming hair at different lengths, making it more lively.

Medium-length hair suits a wide range of people, especially those with oval and long faces. The oval face is quite standard, allowing for various hairstyles, while medium-length hair can further soften facial features. For individuals with

© The Author(s), under exclusive license to Springer Nature Switzerland AG 2025

S. Li, *Haircutting Robots*, https://doi.org/10.1007/978-3-031-84026-5_4

Fig. 1 A hairstylist designs a short haircut for a client in a modern salon, emphasizing layers and freshness

long faces, wavy medium-length hair can balance facial length and add horizontal visual effects, reducing the overly elongated impression [2].

Long hair has always symbolized romance and femininity, available in various forms such as straight, curly, and wavy hair. Straight hair embodies a clean, pure beauty, suitable for both everyday and formal occasions, while curly and wavy hair showcases rich femininity, conveying softness and gentleness.

Long hair is more suited for women with well-defined facial structures, such as square and heart-shaped faces. Naturally falling long hair can soften facial angles, making overall lines appear gentler. Many women choose long hair as their everyday style, not only for its versatility but also for the elegant and romantic image it conveys. For special occasions, long hair can be styled dramatically, showcasing personal beauty and confidence [3].

As fashion evolves, special hairstyles have become important means of expressing individuality and artistry. Techniques like dyeing, braiding, and punk styles enrich hairstyle diversity, allowing people to showcase their creativity and unique styles. Dyeing involves various color combinations, from natural hues to neon colors, offering limitless possibilities for hairstyles. Braiding, with its intricate weaving techniques, creates distinctive hairstyles often seen at weddings and formal events. Punk hairstyles emphasize individuality and avant-garde attitudes, commonly associated with music festivals and art exhibitions.

These special hairstyles cater to individuals seeking personalized expression and strong artistic flair, particularly in settings where showcasing personal style is essential. Studies indicate that hairstyles, as a form of self-expression, can significantly influence others' perceptions of an individual's personality and creativity [4].

2 Principles and Methods of Hairstyle Design

One of the primary principles of hairstyle design is analyzing the client's facial contours. Face shape classifications typically include round, oval, heart, long, and square faces. Each face shape has unique contour characteristics, so hairstyle design must be adjusted according to the structural features of the face. For instance, long faces are characterized by their length, making wavy hairstyles more suitable as they add horizontal width to balance facial proportions. Conversely, individuals with round faces can enhance facial dimension through more layered hairstyles, avoiding overly round visual effects.

When choosing a hairstyle, barbers should focus on how the hairstyle can correct facial asymmetries or prominent features. Research indicates that an appropriate hairstyle can significantly improve the overall balance of the face, making the contours appear more harmonious [5]. This is why each face shape has specific hairstyle pairing principles; selecting the right hairstyle can effectively enhance one's appearance (Fig. 2).

In addition to face shape, hair texture is also a crucial factor affecting hairstyle design. Hair textures can be categorized as fine, coarse, curly, or straight, each yielding different effects in hairstyle presentation. Fine hair typically lacks volume and tends to fall flat, making it suitable for designing fuller hairstyles, such as curls or layers, which increase thickness and dimension to avoid a sparse appearance. Coarse hair, being strong in texture, is ideal for short or structured hairstyles that present a sharp style.

When designing hairstyles, curly and straight hair also require different considerations. Curly hair naturally offers curves, making it easy to create distinct layered effects, while straight hair allows for a clean, sleek line. Understanding the characteristics of hair texture and selecting corresponding hairstyle design methods can help clients maximize the natural beauty of their hair [6] (Fig. 3).

Everyone's temperament and style are unique, so barbers must consider clients' personality traits when designing hairstyles. For instance, individuals with an elegant temperament typically favor softer hairstyles, such as natural curls or medium-length hair, while those pursuing a more avant-garde style may lean towards bold coloring or sharply defined hairstyles.

Hairstyles are not merely aesthetic changes; they can profoundly impact overall image. A hairstyle change often boosts personal confidence, especially in various

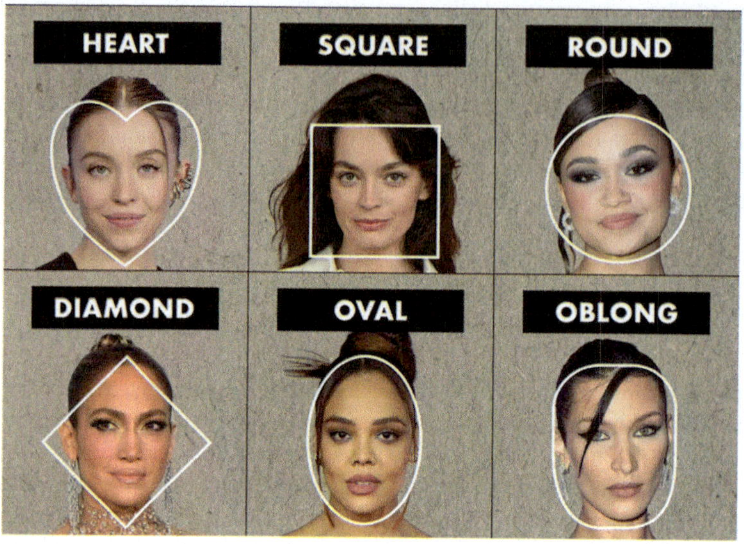

Fig. 2 Different face shapes and their suitable hairstyles, aiding in hairstyle design based on facial contours (https://www.byrdie.com/is-your-face-round-square-long-heart-or-oval-shaped-345761)

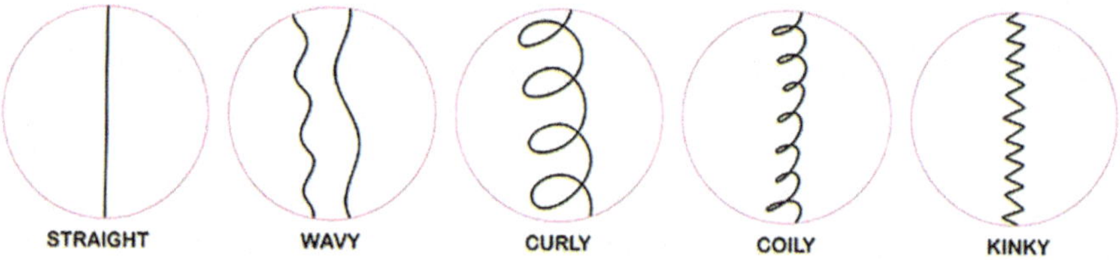

Fig. 3 Examples of different hairstyles

social settings, where a hairstyle that aligns with one's temperament can enhance the overall impression [7]. Therefore, designers should consider clients' lifestyles and professional backgrounds when selecting hairstyles, ensuring that the hairstyle aligns with their overall style.

Hairstyle design should also take into account clients' lifestyles and professional needs. Different professions and lifestyles have varying hairstyle requirements; for example, corporate managers typically require simple and elegant hairstyles to project a professional and composed image, while those in creative industries may prefer uniquely designed hairstyles that express their distinct aesthetic.

Moreover, the ease of managing a hairstyle is an important consideration. For individuals with fast-paced lives, manageable hairstyles are especially crucial, while those with more time and energy can opt for more complex or stylish designs. When designing hairstyles, barbers should find a balance between convenience and fashion, ensuring that the hairstyle is both aesthetically pleasing and suitable for daily life [8].

3 Commonalities and Differences Among Different Hairstyles

Regardless of how diverse hairstyles may be, many commonalities exist in styling principles. A notable example is that layered hairstyles enhance volume by thinning the hair. Whether short or long, introducing layers can make hair appear fuller and avoid the heaviness caused by a single length. Thinning can not only reduce bulk but also increase the airy feel between strands, adding liveliness to the hairstyle. This technique is widely applied in many classic hairstyles.

Another commonality in hairstyle design is the use of bangs to modify the forehead and adjust facial proportions. Bangs can effectively cover a high forehead or elongate a short face, achieving a flattering effect. Whether blunt, side-swept, or wispy bangs, their use serves not only as a decorative element but also as a means of adjusting visual effects on the face [9] (Fig. 4).

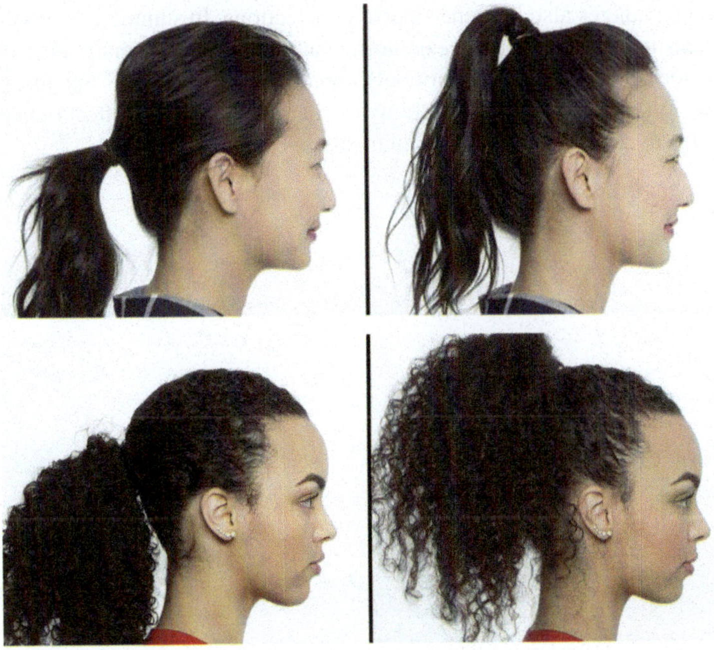

Fig. 4 Comparison of low and high ponytail hairstyles, showcasing style differences for various occasions (https://www.cosmopolitan.com/style-beauty/beauty/a59400/this-hair-trick-will-instantly-change-your-face/)

While many commonalities exist, significant differences arise in specific styling techniques for different hairstyles. For instance, straight and curly hair are styled quite differently. The focus for straight hair is smoothness; maintaining the flow and line is crucial, requiring specialized straightening techniques to avoid split ends or dryness. Conversely, curling involves using curling irons or perms to create curves and waves, with the key being to ensure that the curls appear natural and last long.

The design differences between low and high ponytails also illustrate the diversity of styling techniques. Low ponytails typically convey a calm, elegant feel, suitable for formal occasions and daily commutes, while high ponytails exude energy and vibrancy, ideal for sports or youthful fashion displays. Although both hairstyles fall under the ponytail category, their style differences are significant due to the placement adjustment [10].

The choice of hairstyle is influenced not only by personal preference but also closely tied to the occasion. Classic hairstyles like bobs and French curls often possess enduring fashion appeal, suitable for various formal and casual settings, and are less likely to go out of style. In contrast, trendy hairstyles, such as unique dye jobs, extremely short cuts, or punk styles, depend more on seasonal fashion trends and typically have a shorter lifespan, often popular in specific fashion circles or occasions. Trendy hairstyles are particularly suited for young individuals seeking personality and style (Fig. 5).

Seasonal changes also influence hairstyle selections. In summer, due to the heat, people tend to opt for short or easy-to-manage hairstyles to maintain freshness, while longer hairstyles in winter not only enhance styling richness but also provide some warmth. This seasonal variability in hairstyle choices reflects changes in people's needs across different environments [11].

Fig. 5 A comparison of summer and winter hairstyle designs, demonstrating choices that adapt to different seasons

References

1. Kennedy, K. (2020). *My natural hair is unprofessional: The impact of Black hairstyles on perceived employment-related characteristics* (Master's thesis, Marquette University).
2. Weerasinghe, H., & Vidanagama, D. (2020, December). Machine learning approach for hairstyle recommendation. In *2020 5th International Conference on Information Technology Research (ICITR)* (pp. 1–4). IEEE.
3. Pergament, D. (1999). It's not just hair: Historical and cultural considerations for an emerging technology. *Chi.-Kent L. Rev., 75*, 41.
4. Sims, J. P., Pirtle, W. L., & Johnson-Arnold, I. (2020). Doing hair, doing race: The influence of hairstyle on racial perception across the US. *Ethnic and Racial Studies, 43*(12), 2099–2119.
5. Leftwich, C. A. J., Powe, N., & Fehlbaum, A. E. (2024). Hair and career: The impact of hair bias on women of color. In *Women of Color and Hair Bias in the Work Environment* (pp. 126–153). IGI Global.
6. Samsudin, S., Ramadhani, R. A., & Sanjaya, A. (2024). Recommendation system for selecting haircut models based on facial shape using the viola-jones method. *Jurnal Teknik Informatika (Jutif), 5*(1), 145–152.

7. Lee, E. J., & Kim, J. S., Study of modern hair styles represented in femme fatale images from mass media. *Kor J Aesthet Cosmetol, 6*(4).
8. Park, Y. M., & Baek, K. J. (2020). Analysis of differences in store choice and hairstyle pursuit behaviors according to lifestyle types of one-person hair salon users. *The Research Journal of the Costume Culture, 28*(2), 229–244.
9. Sunhem, W., & Pasupa, K. (2016, February). An approach to face shape classification for hairstyle recommendation. In 2016 *Eighth International Conference on Advanced Computational Intelligence* (ICACI) (pp. 390–394). IEEE.
10. Harrison, S., & Sinclair, R. (2003). Hair colouring, permanent styling and hair structure. *Journal of Cosmetic Dermatology, 2*(3–4), 180–185.
11. Lynch, A., & Strauss, M. (2007). *Changing fashion: a critical introduction to trend analysis and cultural meaning*. Berg.

Core Ideas in the Haircutting Process 5

1 Coordination of Hairstyle and Facial Features

A core task of hairstyle design is to balance facial proportions through adjustments in the hairstyle. Different facial features, such as a wide forehead, wide jaw, or long face, can be modified through appropriate hairstyle design. For instance, individuals with wide foreheads may benefit from hairstyles with bangs, such as blunt or side-swept bangs, which can cover part of the forehead and shorten the perceived length of the face. Those with a wide jaw can opt for hairstyles with waves or curls, as these can soften the angularity of the jaw and create a more harmonious facial line.

Moreover, hairstyle design can adjust the overall visual proportions of the face. For individuals with long faces, side-parted hairstyles or curls can increase horizontal visual width, shortening the vertical length of the face. Conversely, for round or wide faces, vertical hairstyles (such as center parts or straight hair) can elongate the face, making it appear slimmer [1]. Through thoughtful hairstyle design, facial proportions become an adjustable factor in the overall appearance (Fig. 1).

Hairstyles not only affect overall visual proportions but also resonate with facial features, highlighting or softening specific attributes. A classic example is how hairline design can significantly influence the appearance of the forehead. By designing the hairline or incorporating suitable bangs, the forehead can be concealed or emphasized, creating different facial contours. For example, individuals with high hairlines can choose bangs to balance facial proportions, while those with lower hairlines might prefer styles that expose the forehead, making the face appear more open.

The treatment of side hair can also accentuate the jawline, particularly for those with defined jawlines. Slightly shorter side hair can enhance facial depth. Furthermore, hairstyles directly impact the visual focal point of facial features. By creating height at the crown or adjusting the length of hair along the sides,

© The Author(s), under exclusive license to Springer Nature Switzerland AG 2025

S. Li, *Haircutting Robots*, https://doi.org/10.1007/978-3-031-84026-5_5

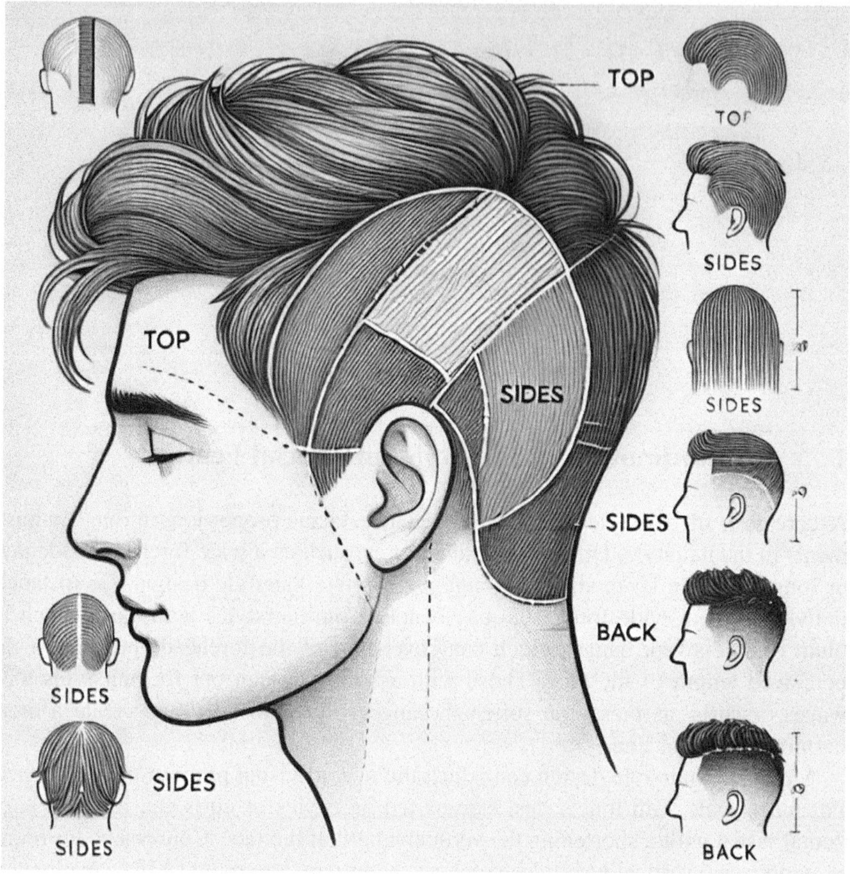

Fig. 1 The impact of different areas of the head on hairstyle design

hairstylists can emphasize clients' eyes or cheekbones, making their features more three-dimensional and pronounced [2].

Hairstyle design is closely related to age, as individuals of different age groups often have varying hairstyle needs. Young people typically favour avant-garde, trendy hairstyles to express individuality and vitality. Such styles are often more dynamic and layered, including popular short cuts, dyed hair, or stylized curls. In contrast, older individuals often prefer simple, elegant hairstyles that reflect maturity and composure. For example, neat medium-length hair or natural waves can beautifully convey elegance and seriousness.

Additionally, hairstyles can help modulate perceptions of age, serving to either reduce age appearance or enhance maturity. Soft curls can make middle-aged women appear younger, while sharp short styles can highlight a polished demeanor, suited for those pursuing a mature image. Designing hairstyles based on age groups not only aligns appearances with life stages but also expresses varying mental states and temperaments [3].

2 Treatment of Lines, Layers, and Texture

The lines of a hairstyle play a critical role in shaping the overall image style. Straight lines typically convey a clean and sharp feel, making such hairstyles suitable for designs that pursue a professional image or modern aesthetic. By using smooth straight hair lines, facial bone structure can be highlighted, presenting a tidy overall appearance. In contrast, curved lines offer a softer look; through natural arcs and waves, they create a warm and elegant visual effect. These styles are often used to embody gentle femininity or romantic flair [4].

Different cutting techniques also affect the presentation of hairstyle lines. Diagonal cutting can create a gradient effect in hair length, adding a sense of movement and layering; blunt cuts are ideal for creating clear hairline definitions, suitable for simple, geometric styles; point cutting thins the hair, resulting in a more natural and lightweight appearance. The flexible application of these cutting techniques enables hairstylists to achieve various line effects based on client preferences, allowing for diverse designs [5].

Layering is another essential element in hairstyle design, significantly enhancing dimensionality and movement. Through cuts at different lengths, hairstylists can make hairstyles appear richer and more varied, avoiding a uniform length. Both short and long hairstyles can gain vitality through layering, making them more lively and interesting.

Layering adapts well to different hair textures. For thick hair, layering can reduce weight, alleviating heaviness and achieving a more lightweight look; for fine hair, adding layers can create more volume, making the hairstyle appear fuller and more abundant [6]. By controlling layering, hairstylists can design personalized hairstyles that cater to different hair textures and face shapes.

The texture of a hairstyle is also a core consideration in haircutting design, as different hair types and textures greatly influence the final effect of the hairstyle. By using suitable cutting techniques and styling tools, hairstylists can achieve various texture effects, from smooth to voluminous, even tousled looks. For instance, smooth hairstyles can be created using sleek straight hair or silky curls, ideal for elegant and formal occasions, while voluminous styles can be achieved through curls or perms, projecting a lively and casual aesthetic.

Furthermore, techniques such as perming, thinning, and dyeing can further enhance the texture of a hairstyle. Perming alters the shape of the hair to increase curliness, making the hairstyle appear fuller; thinning reduces volume, creating a more lightweight and breathable effect; dyeing not only adds color layers but also enhances shine and texture [7]. The combination of these techniques offers endless possibilities for shaping hairstyle texture.

3 The Relationship Between Head Structure and Hairstyle Design

The shape of the head is a foundational consideration in hairstyle design, as different regions of the head, such as the top, sides, and back of the skull, directly influence the presentation of the hairstyle. For instance, for individuals with flatter tops, increasing the height at the crown can create a fuller style, enhancing dimensionality and making the head appear more balanced and rounded. For narrower side profiles, increasing the width of the hair, such as extending and curling the side strands, can broaden the visual impression of the head and balance its proportions [8] (Fig. 2).

Asymmetrical head shapes can also be enhanced through hairstyle design. For flat or asymmetrical areas, techniques in hairstyle design can be employed to conceal them, such as using side-parted styles or selectively adding volume to achieve a balanced visual effect. This modification not only increases aesthetic appeal but also enhances the overall coherence of the hairstyle [9].

Hairstyle design should not only consider the hair's shape but also incorporate the contours of the skull to form a streamlined appearance. The shape and curves of the skull determine the fundamental outline of the hairstyle. By observing the shape of the client's skull, hairstylists can appropriately arrange the layers and flow of the hairstyle, allowing for a natural transition with the head. For example, for clients with flatter tops, heightening the crown or increasing layering can add richness to the upper area, balancing the overall head shape (Fig. 3).

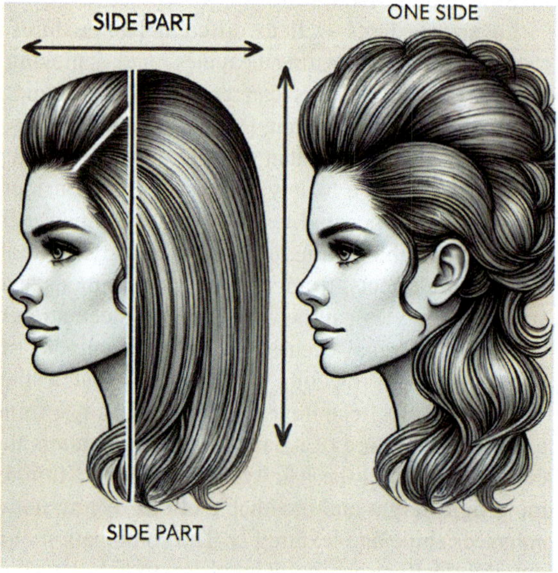

Fig. 2 Illustration of hairstyle modifications for asymmetrical head shapes

Fig. 3 Layered hairstyles balancing flat head structures

In some cases, the contours of the skull significantly impact the layering and height design of the hairstyle. For prominent areas of the skull, shortening or thinning the hair can minimize their visual effect, while for flatter regions, adding volume can fill in and elevate the appearance [10]. Through this method, the hairstyle interacts with the skull's contours, achieving balance and beautifying the head structure.

Hairstyles not only affect appearance but can also influence the visual effect of head posture. By thoughtfully designing the distribution and form of weight in a hairstyle, hairstylists can adjust the balance of the head within the overall image. For instance, if the weight of the hairstyle is concentrated on one side, it may create a tilted or asymmetrical appearance. In such cases, adjusting the hairstyle on the other side can balance the weight, making the head posture appear more natural. Similarly, taller hairstyles can project confidence, while low or overly flat styles may come across as timid (Fig. 4).

Another important consideration in hairstyle design is how the styling can influence the visual height and width of the head. For clients with shorter statures or smaller heads, increasing the height and layering of the hairstyle can effectively elongate the visual length of the head, enhancing overall proportions [11]. This coordination with head posture can significantly improve the overall image, allowing the hairstyle to blend harmoniously with the individual's temperament.

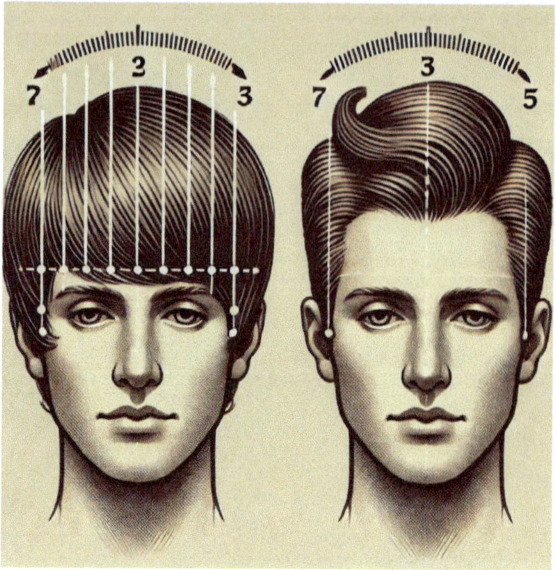

Fig. 4 The impact of hairstyle weight distribution on head posture

References

1. Anjyo, K. I., Usami, Y., & Kurihara, T. (1992, July). A simple method for extracting the natural beauty of hair. In *Proceedings of the 19th annual conference on Computer graphics and interactive techniques*, (pp. 111–120).
2. Robinson, C. L. (2011). Hair as race: Why "good hair" may be bad for Black females. *Howard Journal of Communications, 22*(4), 358–376.
3. Wolfram, L. J. (2001). *Hair Cosmetics*. In Handbook of Cosmetic Science and Technology (pp. 597–620). CRC Press.
4. Pasupa, K., Sunhem, W., & Loo, C. K. (2019). A hybrid approach to building face shape classifier for hairstyle recommender system. *Expert Systems with Applications, 120*, 14–32.
5. Baker, L. M. (1977). Haircutting Guide for Cosmetology Students.
6. Krausel, T., & Rust, R. C. (2015). Hair styling: Technology and formulations. *Cosmetic Dermatology: Products and Procedures*, pp. 270–279.
7. Da Costa, D. (2007). *Textured Tresses: The Ultimate Guide to Maintaining and Styling Natural Hair*. Simon and Schuster.
8. Vetica, R. (2009). *Good to Great Hair: Celebrity Hairstyling Techniques Made Simple*. Fair Winds Press.
9. You, O. L., & Paik, S. Y. Study on post-modernism in modern hair style. *Kor J Aesthet Cosmetol*, *6*(3).
10. Davies, B. J. (2014). *Vintage Hair Styles of the 1940s: A Practical Guide*. Crowood.
11. Harrison, S., & Sinclair, R. (2003). Hair colouring, permanent styling and hair structure. *Journal of Cosmetic Dermatology, 2*(3–4), 180–185.

Analysis of Existing Haircutting Equipment

6

1 Types of Scissors and Their Functions

Scissors are the most fundamental tools in haircutting, with different types allowing for a variety of cutting effects to meet various hairstyle design needs. Barbers must choose the appropriate scissors based on the client's hair type, volume, and desired hairstyle. Common types of scissors include straight scissors, thinning shears, and curved scissors, each serving unique functions and providing different technical support for hairstyle design.

Straight scissors are one of the most commonly used tools by barbers, suitable for a wide range of straight cuts and often used for trimming hair length and contours. The blades of these scissors are typically designed to be straight and sharp, allowing for precise trimming, making them ideal for neat and clear hairstyles, such as classic straight cuts or clean short styles. Due to their straightforward operation, straight scissors are an entry-level tool for novice barbers and a core tool for advanced stylists when creating detailed hairstyles. They are also commonly used for the initial outline of a hairstyle, laying the foundation for subsequent styling [1].

Thinning shears, also known as texturizing scissors, feature serrated blades primarily used to reduce hair volume and add layers. By selectively cutting some strands of hair, thinning shears can effectively decrease overall volume, preventing a heavy hairstyle. For clients with excessively thick hair, thinning shears are an invaluable tool, allowing the hair to appear lighter and more natural without altering the overall structure of the hairstyle. Additionally, thinning shears can break up overly uniform lines, providing a softer layering effect that makes hairstyles look more dynamic and fluid. They are particularly useful for long and thick hair, as they help the hair better adapt to the shape of the head and avoid stiffness [2].

Curved scissors have blades that are curved, mainly used for detailing and shaping hair. They are suitable for curved hairstyles or curved designs, such as trimming bangs, refining sideburns, or shaping the nape of the neck. The curved

© The Author(s), under exclusive license to Springer Nature Switzerland AG 2025

S. Li, *Haircutting Robots*, https://doi.org/10.1007/978-3-031-84026-5_6

blade design allows for following the natural contours of the head, making the curves of the hairstyle appear smoother, especially effective in women's curly or wavy hairstyles. Compared to straight scissors, curved scissors are better suited for detailed work, helping barbers gain better control and precision when handling complex curves and localized areas [3].

Each type of scissors plays a different role in the haircutting process, and by effectively utilizing these tools, barbers can meet the diverse hairstyle needs of their clients, creating unique styles that align with individual preferences and facial features.

2 Structure and Usage of Electric Clippers

Electric clippers are an indispensable part of modern haircutting equipment, especially effective and convenient for short hair and men's hairstyles. Due to their ease of use and quick cutting capabilities, electric clippers are widely employed in salons and home haircutting scenarios. They not only reduce haircut time but also enhance consistency and precision in hairstyling. The core structure of electric clippers includes the clipper blade, electric motor, adjustment mechanism, and handle, all working together to achieve efficient haircutting.

The blade of the clipper is the central component, with sharp blades driven by an electric motor to move rapidly for cutting hair. The design of the blades typically considers different hair types and lengths, allowing for adjustable blade lengths based on needs. Common clipper blade sizes include 3 mm, 6 mm, and 9 mm, allowing barbers to flexibly choose the blade length based on client hairstyle requirements. For clients needing very short hairstyles, barbers usually use the shortest blade to ensure a clean cut, while for those wanting to retain some length, a longer blade is selected [4] (Fig. 1).

The clipper blade may feature self-cleaning or rust-resistant designs to maintain sharpness during frequent use. Some advanced models come equipped with adjustable blade angles, enabling barbers to handle different angles of hairstyle design flexibly and ensure trimming in hard-to-reach areas, such as behind the ears and along the hairline. The sharpness and durability of the blades directly influence the precision and efficiency of the haircut, making the material and design of the blades a primary consideration when purchasing clippers [5].

The electric motor powers the device and is the core driving component of the clipper. Common types of electric motors include rotary motors and vibration motors. Rotary motors are typically used in professional haircutting equipment; they provide powerful performance, capable of efficient operation over extended periods, making them suitable for frequent use by barbers. This type of motor excels at handling various hair types, especially thick and coarse hair, and is less prone to overheating.

Vibration motors have relatively lower power and are mainly used in home clippers. Their advantages include low noise, affordability, and simplicity of operation, making them ideal for everyday home haircutting needs. Although vibration

Fig. 1 Structure and usage of electric clippers

motors have lower power, they are sufficient for typical home use, and their lightweight and convenience make them comfortable to use. Additionally, some high-end clippers utilize brushless motor technology, which enhances the lifespan of the clippers and significantly reduces maintenance frequency [6].

When using electric clippers, barbers should choose the appropriate clipper length based on the client's hairstyle requirements. First, they need to determine the design outline and the desired length of hair to retain. Generally, barbers will start with a longer blade and gradually adjust to shorter blades to ensure the hairstyle's layering and evenness. The blade should align with the direction of hair growth to ensure smooth and natural cutting.

Barbers often use combs in conjunction with clippers for auxiliary trimming. Combs help barbers smooth out the hair and control the length being cut by the clippers, ensuring even trimming. For more precise results, clippers can also be used alongside other haircutting tools in the salon, such as scissors and razors, especially when refining the hairline, sideburns, and areas behind the ears.

Electric clippers also require regular cleaning and maintenance to ensure they remain in good condition over time. After use, barbers should remove any hair residues from the clipper blades and apply special lubricating oil to maintain sharpness and smooth operation. Good maintenance not only extends the lifespan of the clippers but also ensures that the cutting results remain precise with each use [7].

The efficiency and versatility of electric clippers make them a core tool in the haircutting industry. Whether for simple home haircutting or complex professional hairstyle design, electric clippers can flexibly meet various needs and ensure satisfying results with every haircut.

3 The Role of Haircutting Auxiliary Equipment

In the haircutting process, various auxiliary devices play crucial roles alongside core tools like scissors and electric clippers. Although these devices may seem unremarkable, they significantly impact the precision, efficiency, and overall experience of clients. Barbers need to master the use of these devices to ensure the smooth progress of the haircutting process and enhance client satisfaction (Fig. 2).

Haircutting capes are one of the essential pieces of equipment in every haircutting session. Their primary function is to prevent cut hair from falling onto the client's clothing and skin, ensuring cleanliness and comfort. Capes are typically made from lightweight, waterproof materials, effectively preventing water droplets and cut hair from adhering to the client. Modern capes prioritize comfort in design, often using breathable materials to avoid discomfort or overheating during extended use.

In addition to the material, the size and shape of the cape design are also crucial. Haircutting capes are usually large enough to cover the client's upper body and shoulders, ensuring hair does not fall inside their clothing. Some high-end capes include non-slip designs or magnetic fasteners for quick adjustment of fit, preventing awkward situations caused by slipping or shifting during cutting. Additionally, capes can be designed in various styles for different use scenarios; for instance, some may have special insulation layers for dyeing or perming to protect against chemical agents or heated equipment [8].

Spray bottles are essential tools used for wetting hair during haircutting. Barbers typically use spray bottles to evenly mist hair before cutting, making it smoother and reducing static and dryness. When hair is wet, it is easier to control and cut, especially for longer hairstyles, ensuring that the strands are neatly aligned and avoiding irregular cut lines.

The design of spray bottles typically considers the evenness of the spray and control over the amount of water. High-quality spray bottles produce a fine mist, avoiding excessive moisture that can affect cutting results. These bottles are usually made from durable plastic or metal, designed for long-term use. Some professional

Fig. 2 The use of a haircutting cape

spray bottles even feature adjustable nozzles, allowing barbers to modify the spray range and intensity according to different hair types and styles [9] (Fig. 3).

Combs are also crucial auxiliary tools in the haircutting process. Different combs, based on their spacing and shape, are suitable for various hair types and styles. Common types of haircutting combs include wide-toothed combs, fine-toothed combs, and cutting combs.

Wide-toothed combs are ideal for managing thick or curly hair. With larger spacing, they can easily detangle complex hairstyles without disrupting the hair's natural curl or causing damage. Fine-toothed combs are used for precision trimming, suitable for short or medium-length hair, helping barbers better control hair alignment and ensuring neat cuts. When using electric clippers, barbers often pair them with cutting combs to guide the hair into the clipper blades, ensuring even and precise trimming [10] (Fig. 4).

The materials and design of combs are also vital. High-quality haircutting combs are often made from anti-static materials to avoid static build up during

Fig. 3 The role of spray bottles in haircutting

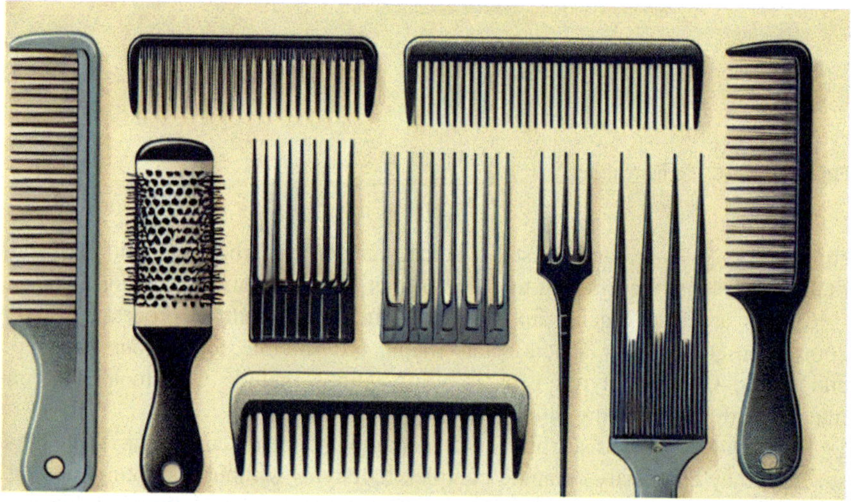

Fig. 4 Different types of combs

combing, which can affect hair management. Additionally, the curvature and flexibility of combs help barbers handle complex hairstyles and the curves of the head more flexibly.

Apart from capes, spray bottles, and combs, other auxiliary equipment such as hair clips, towels, razors, and blow dryers also play significant roles in the haircutting process. Hair clips are used to secure excess hair, allowing barbers to focus on specific areas; towels are employed to clean up cut hair and moisture, ensuring client comfort. Blow dryers are used after cutting to quickly dry hair, ensuring that the final hairstyle appears natural and lasting [11].

These auxiliary devices work together to enhance the efficiency and quality of haircutting, improving the client's experience and making the overall process smoother and more comfortable.

4 Automation in Haircutting Equipment

With advancements in technology, automated haircutting equipment is gradually entering the haircutting industry. Although they currently cannot fully replace barbers, these devices have demonstrated significant advantages in specific scenarios. Existing automated haircutting equipment primarily focuses on efficient, standardized hair trimming, particularly in men's haircuts, children's haircuts, and simple hairstyle designs.

Smart electric clippers are among the most common types of automated haircutting equipment. Compared to traditional electric clippers, smart electric clippers come equipped with automatically adjustable blades and built-in sensors that can adapt the cutting length based on hair volume and texture. This type of device is especially suitable for home use or for quickly managing short hairstyles, with many brands offering present hairstyle modes for users to choose different haircutting plans. The convenience of smart electric clippers allows users to perform haircuts with minimal effort [12] (Fig. 5).

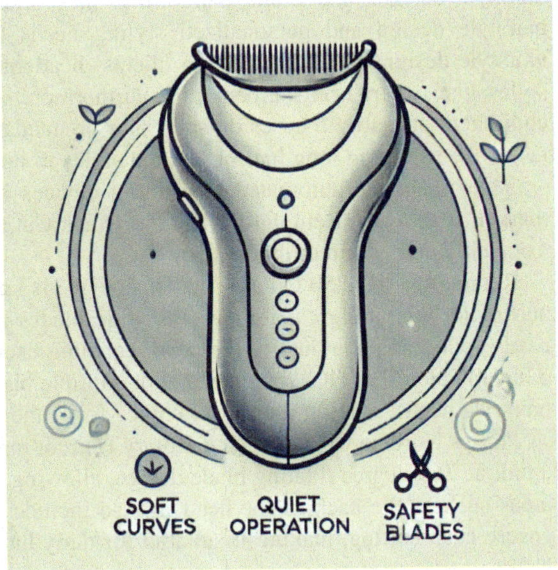

Fig. 5 Special design of children's haircutting equipment

Fully automated haircutting devices are typically used in children's haircuts or self-service home haircuts. These devices use suction or rotating blades to automatically trim hair. Users only need to place the device on their head, and it will automatically detect the hair length and evenly cut off the excess. These devices are designed to provide a simple and safe haircutting experience, especially popular in children's haircuts, helping to alleviate fears children may have due to noise or barber techniques [13].

Automated haircutting devices can greatly improve efficiency, particularly in simple hairstyles, where the devices can quickly complete cuts, reducing the time barbers spend on manual operations. For example, smart electric clippers can automatically adjust the blade based on hair length, ensuring consistent trimming without the need for repeated adjustments [14].

Compared to the manual operations of traditional barbers, automated haircutting devices can offer more consistent results. Whether using preset hairstyle modes or user-customized options, these devices can perform cuts strictly according to settings, avoiding inconsistencies that may arise from human operation. This consistency is particularly important for clients who require regular haircuts, such as men or children who visit the barber frequently [15].

Automated haircutting devices are typically designed with safety in mind, making them suitable for users without haircutting experience. For instance, many devices feature blunt blades or protective mechanisms to ensure that the scalp remains unharmed during cutting. For children or sensitive clients, some devices incorporate silent technology to reduce noise interference during the haircutting process [16].

While automated haircutting devices have significant advantages in simple hairstyle designs and home haircutting, they still face limitations in complex hairstyle design and personalized styling needs. These devices cannot adjust hairstyle designs flexibly based on clients' head shapes, hair types, and personal styles like experienced barbers can. Furthermore, current automated haircutting equipment primarily focuses on short hair or standard hairstyles, leaving complex tasks like trimming long hair or layered styles to manual execution [17].

Several notable automated haircutting devices are currently available on the market, such as the Remington HC4250 Quick Cut and the Philips Norelco Do-It-Yourself Rotary Haircutting System.

Remington HC4250 Quick Cut: This device is known for its convenient design and rapid haircutting capabilities. Its wide blades and curved design allow it to easily follow the contours of the head, enabling users to complete haircuts in just a few minutes. The HC4250 comes with multiple blade lengths to cater to different hairstyle needs, making it an ideal choice for home haircutting [18].

Philips Norelco Do-It-Yourself Rotary Haircutting System: This device features a unique 360-degree rotating blade design, allowing users to easily handle difficult areas such as the back of the head. It also includes smart adjustment features to ensure even cutting, making it suitable for daily haircuts and quick styling needs [19].

The reviewed haircutting equipment highlights three innovative solutions to simplify and enhance the experience of at-home haircuts, each catering to specific user needs while presenting unique advantages and limitations. The Flow bee system, a device combining a vacuum cleaner with a clipper, represents a practical approach to cleanliness and convenience. By using suction to pull hair straight before cutting and collecting the trimmed hair directly into the vacuum, this system eliminates the need for post-haircut cleaning, making it highly suitable for individuals prioritizing tidiness. However, its functionality is primarily limited to simple, uniform haircuts, lacking the ability to accommodate intricate hairstyling or layered designs. Additionally, the requirement for a connected vacuum cleaner introduces potential noise issues, which may detract from its appeal in quieter settings [20] (Fig. 6).

On the other hand, electric clippers equipped with adjustable guard combs provide a more versatile solution, allowing users to achieve a range of hair lengths by swapping out the comb attachments. This flexibility makes them a popular choice for families or individuals seeking the capability to create multiple styles with a single device. Moreover, their straightforward operation supports accessibility for non-professionals. However, users aiming for more complex or sophisticated hairstyles may find these tools require a higher degree of skill. Furthermore, unlike the Flowbee, these clippers lack integrated cleaning functionality, meaning that users must manually handle the loose hair after cutting [21].

Lastly, the UFO-shaped self-haircutting device takes a unique approach, specifically designed for self-use. Its compact and portable design enhances its convenience, making it an appealing option for users who prioritize mobility or wish to maintain simple short hairstyles, such as buzz cuts, at their convenience. Despite these benefits, the device's practical application remains limited to basic cuts, with significant challenges in achieving a wide variety of hairstyles or precise detailing. User feedback also suggests that the results can be inconsistent, often requiring practice or multiple attempts to achieve the desired outcome [22].

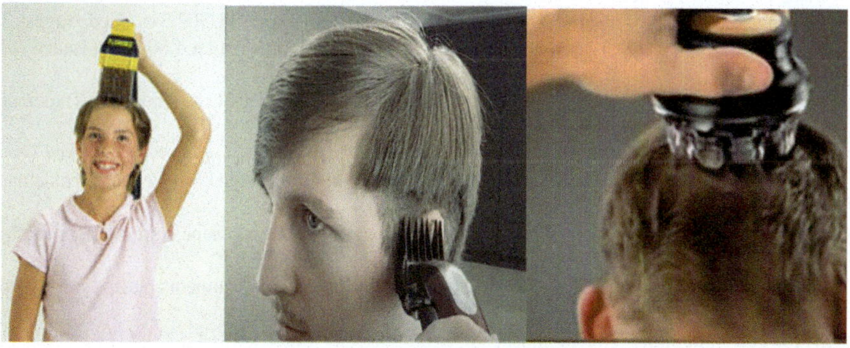

Fig. 6 Different types of haircutting equipment: Flowbee (left), electric clipper with adjustable guard combs (middle), UFO-shaped self-haircutting device (right)

Overall, these devices reflect varying levels of innovation, from prioritizing cleanliness and ease of use with the Flowbee, to enhancing versatility with adjustable clippers, and promoting self-sufficiency with the UFO device. While each tool meets the needs of specific audiences, their limitations underscore the trade-offs between convenience, flexibility, and functionality in the realm of personal haircutting equipment.

References

1. Pike, A. E. (1992). *U.S. Patent No. 5,157,870*. Washington, DC: U.S. Patent and Trademark Office.
2. Davis, M. G., Thomas, J. H., Van de Velde, S., Boissy, Y., Dawson, T. L., Jr., Iveson, R., & Sutton, K. (2011). A novel cosmetic approach to treat thinning hair. *British Journal of Dermatology, 165*(s3), 24–30.
3. Boyles, J. L., Yearout, R. D., & Rys, M. J. (2003). Ergonomic scissors for hairdressing. *International Journal of Industrial Ergonomics, 32*(3), 199–207.
4. Buck Jr, R. N., & Melton, S. (2017). *U.S. Patent No. 9,545,729*. Washington, DC: U.S. Patent and Trademark Office.
5. Li, L. (2022) *Electric hair clipper driven by a brushless external rotor motor*. U.S. Patent Application 17/354,080, filed October 27
6. Nana, B., Yamgoue, S. B., Tchitnga, R., & Woafo, P. (2018). On the modelling of the dynamics of electrical hair clippers. *Chaos Solitons & Fractals, 112*, 14–23.
7. Kammogne, A. S. T., Siewe Siewe, M., Nana, B., Koh, R. E., & Fotsin, H. B. (2024). Dynamic responses of electrical hair clippers with fractional damping and its robust stabilization design. *Journal of Vibration Engineering & Technologies, 12*(1), 1089–1103.
8. Baker, L. M. (1977). Haircutting guide for cosmetology students.
9. Mercurio, A. (1994). *U.S. Patent No. 5,344,076*. Washington, DC: U.S. Patent and Trademark Office.
10. Raskin, R. (2014) *Apparatus and method for cutting hair*. U.S. Patent Application 13/829,189, filed September 18
11. Emmett, F. E., & Luzaich, G. (1996). *U.S. Patent No. 5,496,090*. Washington, DC: U.S. Patent and Trademark Office.
12. Matthews, M., & Harvin, G. (2018). *U.S. Patent No. 9,914,228*. Washington, DC: U.S. Patent and Trademark Office.
13. Cheng, I. (2007) *Safety haircut scissors*. U.S. Patent Application 11/307,246, filed August 2.
14. Natrasevschi, A. (1986) *Automatic hair cutting apparatus*. U.S. Patent 4,602,542, issued July 29.
15. Krenik, M. W. (2017). *U.S. Patent No. 9,656,400*. Washington, DC: U.S. Patent and Trademark Office.
16. Ringart, Y. N., Gamliel, J., & Schaeffer, G. W. (2009). *U.S. Patent Application No. 12/077,883*.
17. Krenik, M. W. (2018). *U.S. Patent No. 9,925,676*. Washington, DC: U.S. Patent and Trademark Office.
18. Remington HC4250 User Manual: https://cdn-img.remington-europe.com/manager/remington-europe_com/User%20Manuals/HC4250_UK_T22-7000573.pdf
19. Philips Norelco Haircutting System User Manual: https://www.documents.philips.com/assets/20220415/6d5362be2d8a425aa52bae7800dd976f.pdf
20. Flowbee Haircutting System https://www.youtube.com/watch?v=nfu4-SZU_E0
21. Electric Clippers with Adjustable Comb Guards https://www.youtube.com/watch?v=eUwR_CbezmQ
22. UFO Self-Haircutting Device https://www.youtube.com/watch?v=GtRBxbgeNIw

Trends in Automation of Haircutting Equipment

7

1 Current Status of Automated Haircutting Equipment

With technological advancements, automated haircutting equipment has gradually become a part of the haircutting industry, providing convenient and efficient solutions for various user groups. The design and functionality of these devices cover everything from everyday home use to professional haircutting needs, meeting application requirements in different scenarios. To gain a deeper understanding of the current state of automated haircutting equipment, we will analyze the classification of devices, technological maturity, and market demand (Fig. 1).

Home smart haircutting devices are currently one of the most common automated tools on the market, primarily targeting average consumers, especially families wishing to cut hair themselves. Compared to traditional manual haircutting tools, home smart devices simplify the haircutting process through automation. Most home smart haircutting devices come equipped with automatic adjustment features that can adapt the clipper blade length according to hair length and type, offering various modes and present hairstyle options. Some devices even integrate smart sensors that can detect hair thickness and density in real-time, automatically adjusting cutting speed to ensure uniform and consistent results.

This type of device is particularly popular among male consumers, as they frequently need haircuts, and home smart haircutting devices offer a quick and precise solution. These devices are usually affordably priced, covering everything from basic length trimming to detailed finishing. Renowned brands like Philips and Remington have seen widespread use of their home smart haircutting devices globally, especially in households looking to save time and haircutting costs.

Professional automated haircutting systems are mainly used in salons, with designs that are more complex and advanced. Unlike home devices, professional automated systems emphasize high precision, multifunctionality, and long operational hours, often equipped with advanced robotic arms, sensors, and 3D scanning technologies. These systems can automatically recognize customers' head shapes

© The Author(s), under exclusive license to Springer Nature Switzerland AG 2025
S. Li, *Haircutting Robots*, https://doi.org/10.1007/978-3-031-84026-5_7

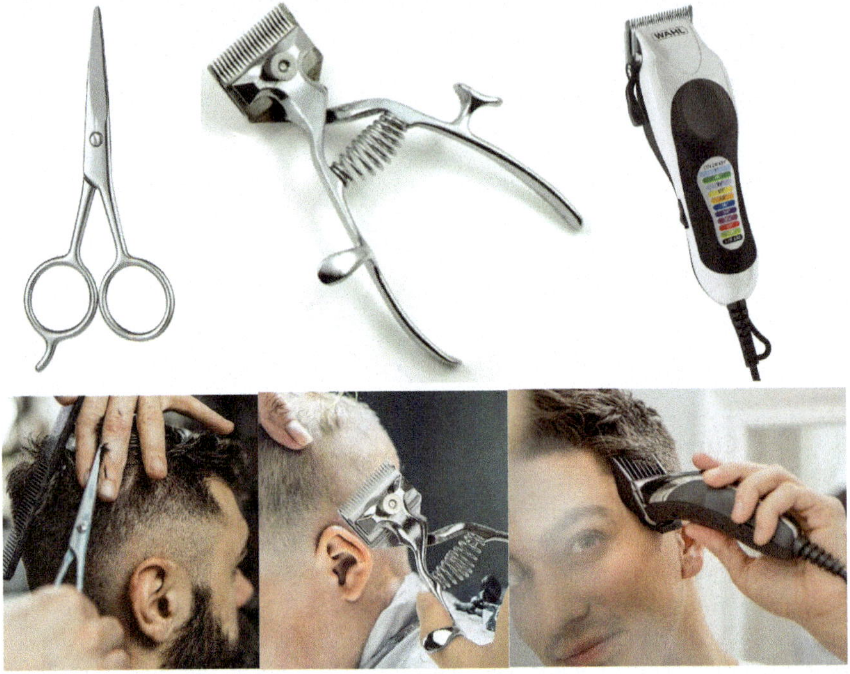

Fig. 1 Evolution of haircutting tools. Early devices used general scissors, later manual clippers were invented, followed by electric clippers, and future innovations may include automated haircutting machines and remote haircutting devices

and hair types, providing personalized hairstyle design and trimming based on customer needs. Their automation capabilities not only improve efficiency but also reduce human error, particularly advantageous when handling numerous customers or complex hairstyles.

Although professional automated systems are relatively expensive, they are popular in high-end salons and chain barber shops. These devices not only enhance haircutting efficiency but also provide standardized hairstyle designs and a better user experience. As artificial intelligence and robotics technology continue to develop, these systems are expected to become more prevalent in the future [1].

Children's haircutting represents a unique market, with many parents wishing to cut their children's hair at home instead of taking them to salons. Since children often feel anxious or uncomfortable during haircuts, automated haircutting devices designed for children prioritize safety, quiet operation, and gentle cutting experiences. These devices typically feature protective designs and gentle blades to ensure no harm comes to children's scalps. Smart functions are also incorporated, such as automatic speed adjustments and silent modes, to minimize children's fear during haircuts.

Some devices even incorporate cartoon designs and entertainment features to help children feel comfortable and enjoy the haircutting process. This type of equipment has stable market demand, especially when parents wish to provide a safe and convenient haircutting experience for their children.

Basic automated clippers are among the most mature and widely adopted haircutting devices on the market. These clippers are commonly used in homes and regular barber shops, meeting standard haircutting needs. Most automated clippers are equipped with adjustable blades and electric motors, allowing users to perform daily haircuts with simple operations. These devices typically have long battery lives and support quick charging, maintaining high efficiency during frequent use.

With advancements in technology, automated clippers have gradually implemented multi-mode adjustments and smart controls, enabling users to select suitable trimming modes based on different hairstyle needs. Especially for high-end models, features such as built-in sensors that detect hair thickness and automatically adjust cutting force are now available. These clippers, with their practicality and relatively low prices, have become the most widely used products in the automated haircutting market.

High-end automated haircutting equipment is primarily used in upscale salons and special occasions, with a relatively high level of technological maturity, although their market share remains small. Due to the high costs of design and manufacturing, these devices are mainly found in chain salons or high-end beauty institutions located in major cities. They can automatically complete complex hairstyle designs, precise trims, and layering processes through highly intelligent control systems, reducing the workload for barbers.

High-end automated devices typically integrate 3D scanning and facial recognition technologies to automatically generate hairstyle design proposals based on customers' head shapes, facial proportions, and hair types, and adjust operations in real-time during cutting. Although their market share is relatively small, the application areas for these devices are expanding, particularly among clients with high demands for hairstyle precision and consistency.

The market demand for automated haircutting equipment is driven by several key factors. Firstly, there is a growing consumer demand for time and efficiency; modern society operates at a fast pace, with many people wishing to reduce wait times through quick haircutting devices. Secondly, as living standards rise, more consumers are willing to invest in higher-end smart haircutting devices, especially in the home and professional sectors. Thirdly, the aging population has increased demand for easy-to-use haircutting devices that require minimal technical skills, particularly for specific groups such as the elderly and disabled.

Additionally, the impact of the global pandemic has also propelled the demand for automated haircutting devices. Many individuals have opted to reduce outings due to hygiene and safety concerns, leading to an increase in home haircutting equipment sales [2].

Regionally, the penetration rate of automated haircutting equipment is higher in developed countries, particularly in North America, Europe, and Japan. Consumers in these regions show greater acceptance of automation and smart products, with robust demand for new technologies. In contrast, the penetration of devices in developing countries is lower; however, with the expansion of the middle class and changing consumer needs, the market potential in these regions is substantial.

In emerging markets like China and India, the large number of consumers and rising living standards are gradually facilitating the entry of automated haircutting devices into homes and regular barber shops. Additionally, some high-end equipment is trying to penetrate high-income groups in these regions, driving further market expansion [3].

2 Development and Application of Smart Haircutting Devices

With the advancement of automation technology and artificial intelligence, smart haircutting devices have become an important branch of the haircutting industry. These devices combine advanced sensor technology, AI, and machine learning algorithms to automatically adjust according to user needs, providing a more personalized and efficient haircutting experience. The following sections will analyze the core technologies, practical application scenarios, and future development potential of smart haircutting devices.

The core of smart haircutting devices lies in their built-in sensors and data collection technology. These sensors can real-time detect hair thickness, length, density, and scalp shape, ensuring that the devices can automatically adapt to users' hair characteristics during operation. The sensors not only monitor the condition of the hair but also collect related data for future haircutting operations. For example, laser and infrared sensors can accurately measure hair strands' lengths, helping devices automatically adjust blade lengths to avoid cutting too short or too long.

The continuous advancement of these sensor technologies allows smart haircutting devices to complete hairstyle designs more efficiently and precisely, significantly enhancing the user's haircutting experience. By providing real-time data feedback, the devices can adjust cutting force and speed, achieving a more personalized and precise cutting effect [4] (Fig. 2).

Artificial intelligence (AI) and machine learning technologies are key drivers of smart haircutting devices' development. AI systems can optimize device workflows by continuously learning users' haircutting habits, hair characteristics, and commonly used hairstyles. Machine learning algorithms can identify cutting patterns for different hairstyles, enabling automated adjustments. For instance, users can select their desired hairstyle via the device's app, and AI will adjust parameters during the cutting process based on the user's head and hair characteristics to ensure the final result meets expectations.

Currently, some high-end smart haircutting devices can utilize AI for automatic design and trimming of complex hairstyles, reducing users' reliance on haircutting skills and achieving results comparable to professional barbers [5]. By combining AI and machine learning, smart haircutting devices not only improve cutting quality but also provide users with a more personalized cutting experience (Fig. 3).

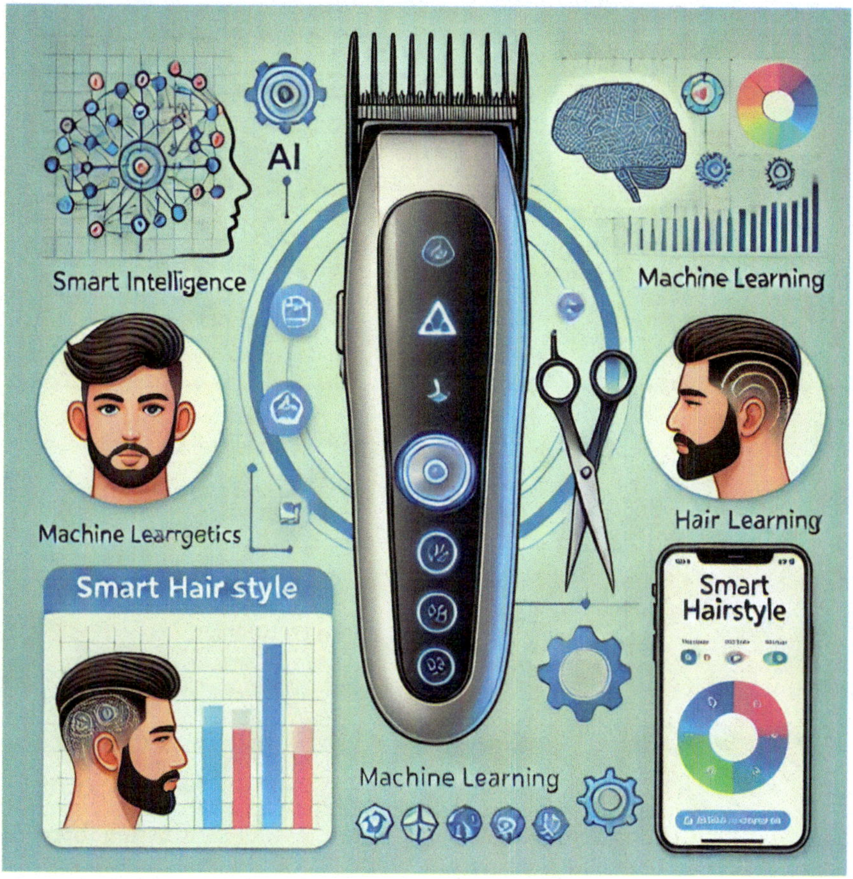

Fig. 2 AI technology empowering the haircutting industry

An automatic adjustment system is another critical feature of smart haircutting devices. This system can automatically adjust blade length, cutting speed, and force according to users' hair conditions and hairstyle requirements. Smart haircutting devices typically come equipped with adaptive systems that adjust cutting force based on hair thickness and density, preventing damage to hair or uneven trimming.

The adaptive cutting system can also monitor changes during the cutting process in real-time; for instance, if the device detects that hair in a certain area is too thick, it will automatically adjust the clipper's angle or speed to ensure even trimming. The combination of automatic adjustment and adaptive features not only enhances haircutting efficiency but also ensures precision and aesthetic appeal in the cutting results [6].

One of the most common applications of smart haircutting devices is home self-haircutting. These devices are designed to enable users to achieve high-quality

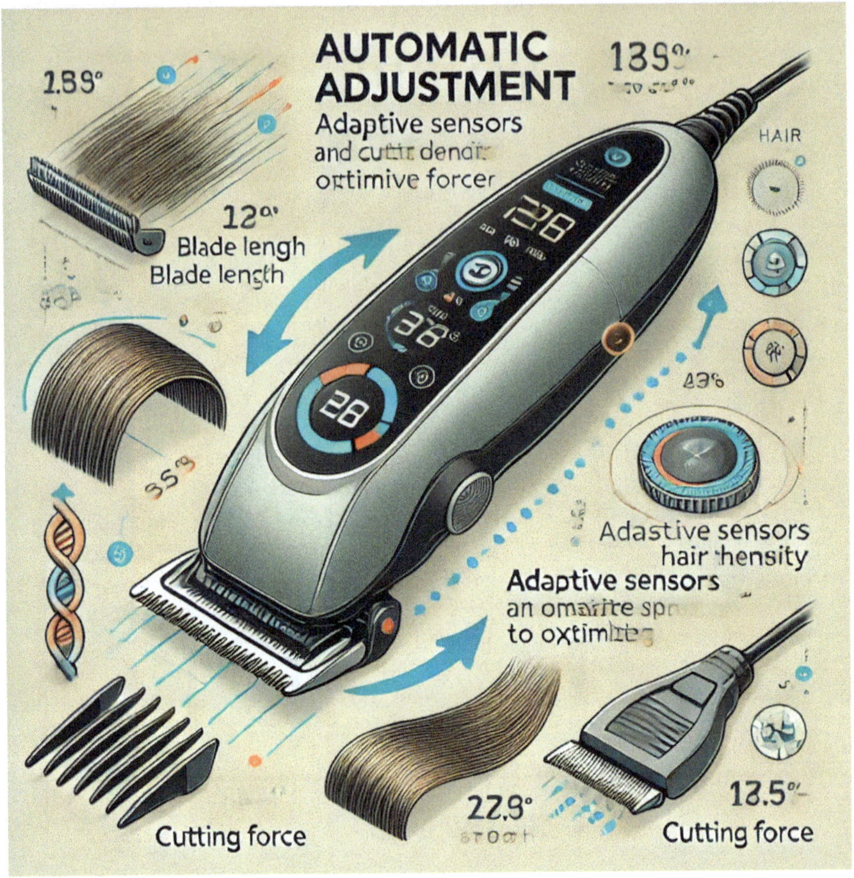

Fig. 3 Automatic control technology adjusts cutting parameters based on specific conditions for customized trimming

haircuts at home without relying on barbers. With built-in smart systems, users need only select their desired hairstyle, and the device will automatically adjust to the appropriate cutting mode, ensuring a simple path to achieving their ideal hairstyle. This home self-haircutting solution is not only convenient and quick but also saves time and costs, particularly appealing to men who require frequent haircuts.

Home smart haircutting devices typically come with multiple blade lengths and different cutting modes, allowing users to adjust the blade length and trimming style as needed. For individuals lacking haircutting skills, smart haircutting devices offer an effective solution, enabling them to easily manage everyday haircuts without a salon visit.

The application of smart haircutting devices is not limited to homes; many salons are also beginning to integrate this new technology to improve efficiency and service quality. In professional environments, smart haircutting devices can help barbers complete hairstyle designs more quickly and accurately, particularly when dealing with complex styles. The AI systems can automatically optimize the cutting process, reducing human error. However, in salon applications, smart devices still face challenges. For example, due to the high cost of equipment, their adoption rate is limited, and the interaction between barbers and clients remains crucial in high-end hairstyle designs, making complete reliance on smart devices not yet mainstream.

For special groups like children and the elderly, smart haircutting devices also demonstrate unique advantages. For children, haircutting is often a process filled with fear and anxiety, and smart haircutters can provide a more comfortable experience through quiet operation and gentle cutting techniques. Some child-specific smart haircutters even incorporate playful designs and entertainment features to further reduce children's fear of haircuts.

For the elderly, smart haircutting devices offer lightweight and easy-to-operate solutions, particularly beneficial for seniors who may have mobility issues or require special care, helping them complete daily haircuts safely and efficiently.

The future development trend of smart haircutting devices undoubtedly points towards a more intelligent and personalized haircutting experience. As AI technology continues to advance, future smart haircutting devices may better understand users' personal hairstyle preferences and provide customized hairstyle solutions based on historical data and real-time analysis. Furthermore, devices may focus more on user feedback and experiences, continuously optimizing cutting results through built-in learning algorithms to deliver highly personalized haircutting services.

Personalized haircutting may not only be achieved through automatic systems; future devices could integrate facial recognition, 3D scanning, and other technologies to offer tailored hairstyle designs based on users' facial features and head shapes, ensuring each haircut meets user needs [7].

With the rapid development of augmented reality (AR) and virtual reality (VR) technologies, future smart haircutting devices may combine these technologies to provide more intuitive hairstyle previews and design functionalities. Users could view real-time simulations of different hairstyles on their heads through the smart haircutting device's application or display screen, allowing them to make more informed decisions before getting a haircut.

Through AR/VR technology, users could see the actual effects of different hairstyles on themselves before cutting, enhancing their haircutting experience and helping them make wiser choices. For barbers, AR/VR technologies could serve as auxiliary tools to better understand clients' needs and provide more precise hairstyle designs.

3 Related Technologies, Patents, and Research Findings

As the trend of automation in haircutting devices continues to deepen, relevant technologies, patents, and research findings are becoming increasingly abundant. Major tech companies and research institutions are dedicated to advancing the intelligence of haircutting devices through sensor technology, AI algorithms, and automated cutting systems, gradually achieving high levels of automation and personalized services. The following sections will detail related technology patents, research advancements, and future technological breakthroughs.

Sensor technology is one of the core components of automated haircutting devices. Through real-time data collection and analysis, it helps devices automatically adjust operational parameters to ensure precise cutting results. Currently, many automated haircutting devices on the market have applied for patents related to sensors. For example, some patents focus on how to use laser or infrared sensors to detect hair thickness, length, and density in real-time, assisting devices in automated adjustments [8].

Another category of patents pertains to cutting mechanisms, typically describing how haircutting devices work in conjunction with sensor technologies to achieve automated cutting operations. For example, some patents outline how devices automatically adjust blade lengths and cutting force based on sensor feedback, ensuring consistent hair lengths and avoiding misoperations. These patents provide technical assurance for the automation of haircutting devices in handling complex hairstyles, helping achieve more refined designs [9].

Intelligent adjustment systems represent another critical technological area within automated haircutting devices. Relevant patents primarily focus on how devices can utilize AI technologies to implement automated adjustment functions. Intelligent adjustment systems often integrate AI algorithms that automatically adjust blade lengths, clipper speeds, and cutting angles based on users' personalized needs and real-time feedback. Some leading patents have demonstrated how intelligent adjustment systems can continually optimize haircutting processes through deep learning algorithms, providing more personalized cutting experiences.

Moreover, some patents also involve how intelligent adjustment systems analyze users' hairstyle preferences and automatically apply those preferences in future haircuts. Patents describe how devices can learn users' haircutting habits and adaptively adjust cutting modes based on different hair types and volumes, ensuring every haircut meets users' expectations. Such technological patents hold significant strategic value in the automated haircutting device market.

Intelligence and human–machine interaction are crucial research topics in the field of automated haircutting devices. Many research institutions and companies are exploring how to achieve seamless interaction between devices and users through intelligent technologies. A significant research direction involves improving human–machine interfaces to enable users to control devices more conveniently, set hairstyle parameters, and make real-time adjustments. For example, through voice recognition technology or touch screens, users can directly instruct devices to perform specific cutting operations (Fig. 4).

Additionally, research focuses on how to leverage augmented reality (AR) technology to allow users to preview hairstyle effects, thereby reducing decision-making time before cutting. With AR technology, users can see simulated effects of different hairstyles before cutting, enhancing their experience and assisting barbers in better meeting customer needs. Such research not only propels the intelligence of haircutting devices but also enhances human–machine interaction experiences.

Globally, numerous research institutions are committed to the technological development of automated haircutting devices. Domestic and international research trends focus on improving the efficiency and user experience of haircutting devices through smart sensors and AI technologies. For instance, some research teams in China are developing low-cost, easy-to-use home smart haircutting devices aimed at popularizing smart haircutting technology and enhancing its usage in households.

In contrast, in Europe and North America, research emphasizes the intelligence and automation capabilities of high-end haircutting equipment. For example,

Fig. 4 Remote haircutting systems utilizing VR technology and robotic arms

some research projects in the United States focus on developing highly automated haircutting robots that achieve complex hairstyle design and trimming through machine learning and AI algorithms. This research plays a vital role in driving innovation and development within the global haircutting device market.

In the future, haircutting robots will undoubtedly become an important development direction for automated haircutting devices. Haircutting robots not only need to possess precise cutting abilities but also require sensor technologies and AI algorithms for self-adjustment and optimization. Potential research directions include how to achieve precise cutting of hair through multi-sensor fusion technology and how to enhance the adaptability of haircutting robots through deep learning algorithms, enabling them to automatically generate hairstyle design proposals based on user preferences (Fig. 5).

Research will also focus on how to enhance the flexibility of haircutting robots through intelligent arms and tactile feedback technologies, allowing them to handle more complex hairstyle trimming. For instance, future haircutting robots may incorporate 3D printing technology to offer more refined hairstyle designs and layering processes, making hairstyle design more personalized and innovative (Fig. 6).

The development of automated haircutting devices will increasingly merge with other technological fields. The combination of artificial intelligence, robotic arm technology, and virtual reality will provide stronger functionalities and broader application scenarios for haircutting devices. For example, the integration of AI and robotic arms will make haircutting devices more flexible and precise when executing complex operations, ensuring that users' hairstyle needs are fully met.

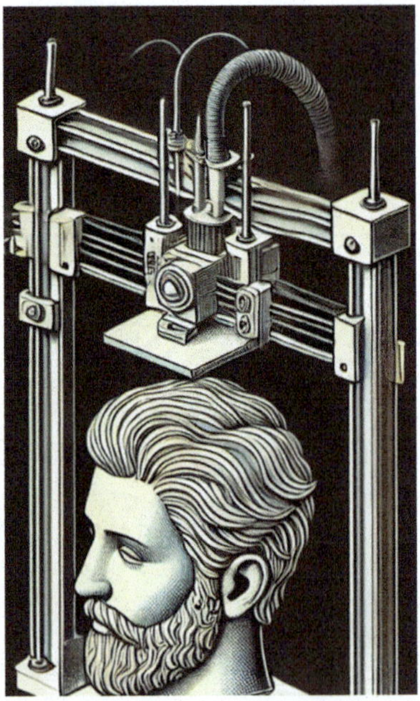

Fig. 5 Automated haircutting devices inspired by 3D printing technology, which can cut excess hair at corresponding positions through the movement of a 3D printer's end effector, enabling complete hair reshaping

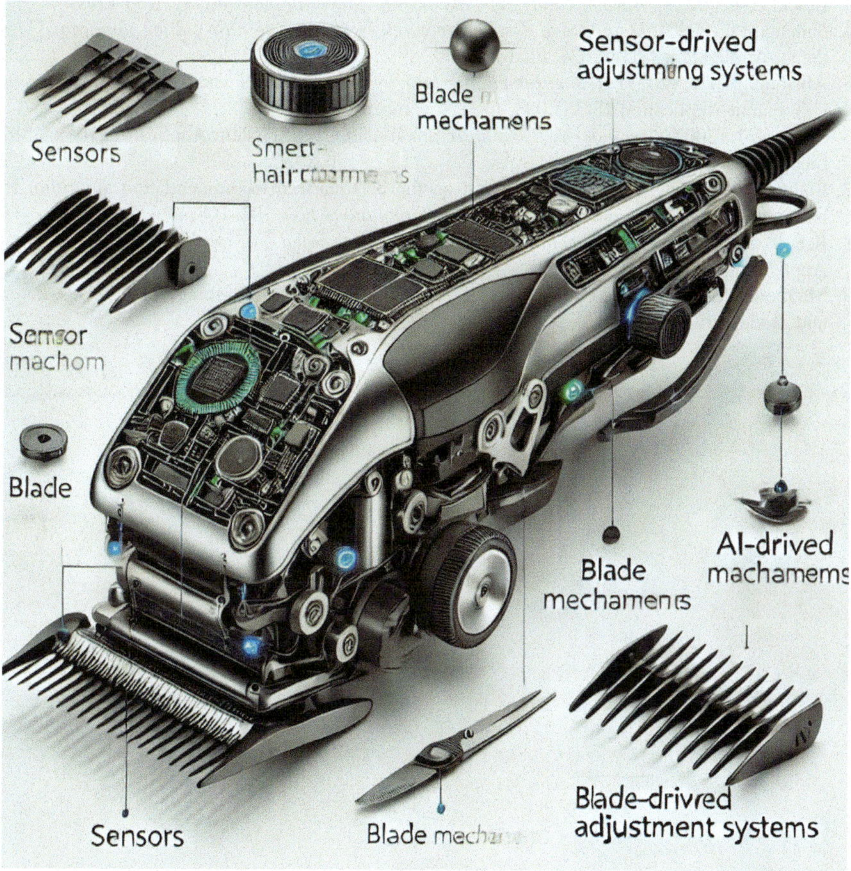

Fig. 6 The future of automated haircutting devices will benefit from a series of advanced technologies, including sensors, high-performance motors, intricate mechanism designs, artificial intelligence, and feedback control technology, fully demonstrating their intelligence

Furthermore, virtual reality technology may play a crucial role in future haircutting devices, allowing users to preview and design hairstyles, confirming effects through virtual experiences before cutting. The fusion of such interdisciplinary technologies will not only advance technological progress in haircutting devices but also significantly enhance user experiences and satisfaction.

References

1. Zandsteeg, C. J., et al. (2022) *System, appliance and method for automated hair processing procedures*. U.S. Patent No. 11,318,630. 3 May.
2. Gerstell, E., Marchessou, S., Schmidt, J., & Spagnuolo, E. (2020). How COVID–19 is changing the world of beauty. *McKinsey & Company*, 8.

3. Block, E. L. (2024). *Beyond vanity: The history and power of hairdressing*. MIT Press.
4. Santana, D. (2022). *Method and system for wireless monitoring hair cutting apparatus*. U.S. Patent Application 17/139,626, filed June 30.
5. Alkahtani, B. Y. (2023) *Helmet for cutting hair by artificial intelligence and method thereof*. U.S. Patent Application 18/331,088, filed November 30.
6. Vitale, J. J. (2002) *Computer-assisted haircutting method*. U.S. Patent Application 10/068,882, filed August 29.
7. Liang, Z., & Wang, F. (2016). Research on the personalized recommendation algorithm for hairdressers. *Open Journal of Modelling and Simulation, 4*(3), 102–108.
8. Krenik, M. W. *Sensing and control techniques for automated hair cutting system*. U.S. Patent Application No. 14/086,497.
9. Matthews, M., & Harvin, G. (2018). *U.S. Patent No. 9,914,228*. Washington, DC: U.S. Patent and Trademark Office.

Transition from Traditional Equipment to Haircutting Robots

8

1 The Need to Transition from Traditional Equipment to Haircutting Robots

With the advancement of society and technology, traditional haircutting methods are increasingly unable to meet the growing demands for personalization and efficiency. Haircutting robots, as a significant transformation direction for the future haircutting industry, can address many limitations of traditional equipment, enhancing efficiency and optimizing the user experience. The following points explain the necessity of transitioning from traditional haircutting equipment to haircutting robots.

1.1 High Skill Requirements in Haircutting

Haircutting is a manual labour that heavily relies on the skills of the hairdresser. There are significant differences in skills and experience among hairdressers, leading to varying quality in haircuts. In high-end salons, skilled hairdressers can provide exquisite hairstyle designs, whereas less skilled hairdressers in regular salons may struggle to meet user expectations. The haircutting process involves complex techniques, such as the angles, force, and layering of haircuts, which require years of experience to master. For those without professional training, self-haircutting is nearly impossible [1] (Fig. 1).

Robotic technology can address these shortcomings. By utilizing sensors, computer vision, and artificial intelligence, haircutting robots can achieve precise identification and processing of hair, providing consistent high-quality haircutting services in various scenarios. Automated devices can adjust based on parameters such as hair length and texture, reducing dependence on hairdresser skills and realizing standardized operations [2].

© The Author(s), under exclusive license to Springer Nature Switzerland AG 2025

S. Li, *Haircutting Robots*, https://doi.org/10.1007/978-3-031-84026-5_8

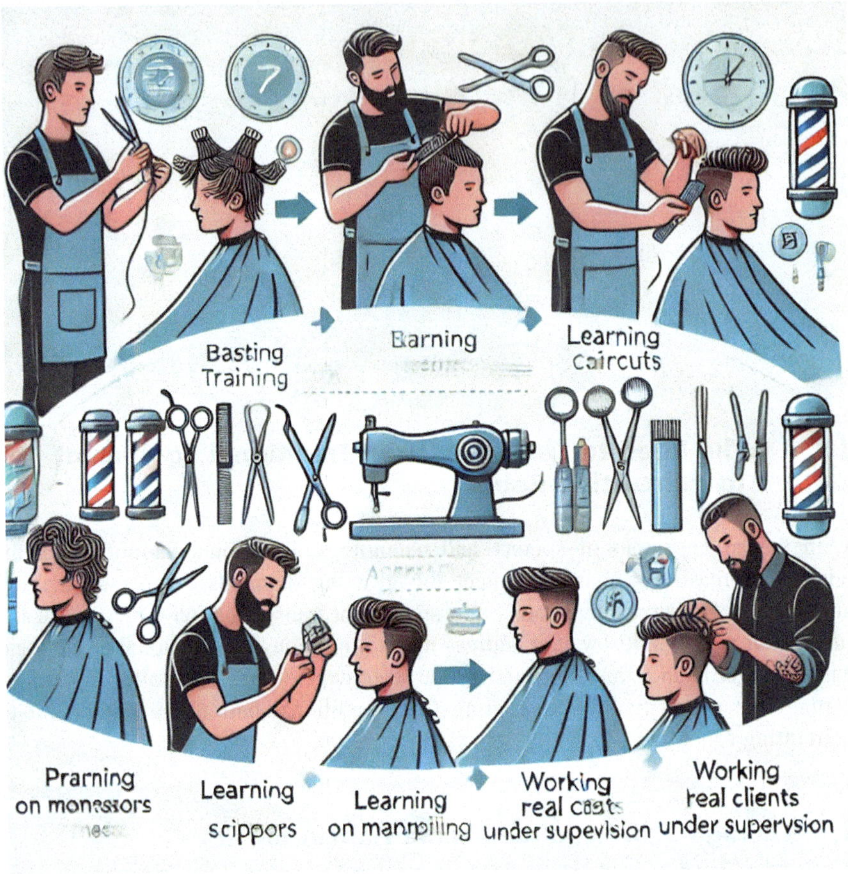

Fig. 1 The learning cycle of a haircutting apprentice from beginner to skilled hairdresser. Different stages show the apprentice's progression from basic training to mastering complex techniques, eventually becoming a professional hairdresser

1.2 High Costs and Supply–Demand Imbalance in Haircutting

Due to the scarcity of skilled hairdressers, the service fees in the haircutting industry are relatively high, especially in high-end salons in large cities where haircut costs can be exorbitant. To reduce expenses, many consumers extend the intervals between haircuts, leading to delayed image management, which adversely affects their personal and professional appearance. Additionally, the number of hairdressers in remote areas is limited, making it difficult for many skilled hairdressers to find enough clients, while those in urban areas face overwhelming demand, creating an imbalance [3] (Fig. 2).

Fig. 2 The high skill requirements in haircutting result in generally high costs

The emergence of robotic haircutting equipment is expected to break this pattern. As technology matures, the costs of haircutting robots are gradually decreasing, making them applicable on a large scale in homes, remote areas, and nursing homes. Haircutting robots not only help consumers reduce costs but also provide services to remote areas through remote operation technology, ensuring resource allocation is optimized.

1.3 Development of Robotic Technology and Cost Reduction

In recent years, advancements in core technologies such as sensors, communications, and direct-drive motors have significantly reduced the costs of robotic systems. The application of sensor technology is particularly crucial in haircutting robots. Through technologies such as obstacle avoidance sensors, distance sensors, and stereo vision, robots can perceive the density, length, and shape of hair in real-time, automatically adjusting their cutting strategies. This not only enhances the safety of the equipment but also significantly improves the accuracy of haircuts [4].

The advancements in direct-drive motor technology provide the basis for haircutting robots to respond quickly. Direct-drive motors not only facilitate rapid and precise cutting operations but also ensure safety for the scalp through force control technology. Compared to traditional motors, direct-drive motors have higher response speeds and lower latency, enabling robotic haircutting devices to handle complex cutting tasks more quickly and accurately [5].

1.4 Future Prospects Driven by Technology

In the future, as robotic technology continues to develop, advancements in force control technology, force feedback technology, and precise mechanism design will further propel haircutting robots toward full automation. Robots could utilize the concept of 3D printing to layer haircuts, leveraging path planning technology and CNC principles to enhance the diversity and precision of hairstyle designs. The automation and intelligence of robotic haircutting equipment will bring revolutionary changes to the industry, gradually transitioning from a supportive role of manual haircutting tools to fully automated haircut solutions [6].

2 Limitations of Traditional Haircutting Equipment

Although traditional haircutting equipment has developed significantly over the past few centuries, its limitations are becoming increasingly apparent in modern society. As consumers' demands for haircut quality, efficiency, and personalization continue to grow, the shortcomings of traditional manual tools are becoming more pronounced.

2.1 Insufficient Efficiency and Consistency

Traditional manual haircutting tools, such as scissors and manual clippers, rely entirely on the skills and experience of the hairdresser. Therefore, the quality and results of each haircut largely depend on the hairdresser's condition and operation on that particular day, making it impossible to guarantee consistent outcomes. This issue is particularly prominent in complex hairstyle designs, which require precise angles, lengths, and layering techniques that are easily affected by human error [7]. Even experienced hairdressers find it challenging to maintain completely consistent results across multiple clients.

In contrast, haircutting robots can ensure consistency and high precision in every operation through the precise control of sensors and algorithms. This standardized operation effectively minimizes human error, providing more reliable and consistent haircut results [8].

2.2 Variability in Skills and Experience

Haircutting is a skilled profession, and there are notable differences in skill levels among hairdressers. Even in the same salon, customers often experience varying haircut quality due to different hairdressers, and even the same hairdresser may perform differently at different times. Due to the limitations of skill levels, many ordinary hairdressers struggle to provide consistently high-quality services in large-scale operations, particularly in complex hairstyle designs [9].

Robotic haircutting equipment can bridge this gap by achieving a highly consistent haircutting experience through programmed and standardized operations. Regardless of when users utilize the service, haircutting robots can execute precise cutting operations according to pre-set programs, reducing the instability caused by skill differences [10].

2.3 Time-Consuming Manual Operations

Manual devices require a longer operating time, especially when performing intricate hairstyle designs, where manual operations often consume a significant amount of time. To ensure accuracy, hairdressers frequently adjust the angle and position of tools while trimming hair in layers, which not only takes time but also increases customer waiting time [11]. Additionally, prolonged manual operation can lead to fatigue for hairdressers, affecting the quality of subsequent services.

Automated haircutting equipment can significantly reduce haircut times. By automatically adjusting the blade length, cutting speed, and force, haircutting robots can complete complex hairstyle designs in a shorter period. This technology not only improves work efficiency but also allows for providing high-quality services to more customers in less time, enhancing the user experience [12].

3 Possibilities and Challenges of Automation Upgrades

In the transition of haircutting equipment from traditional to modern automation, the application of automation technology offers unlimited possibilities for improving haircut efficiency, precision, and user experience. However, this upgrade process also comes with many technical challenges, particularly in key areas such as sensors, communications, speed, safety, force control, and feedback technology. The following sections will explore these possibilities and challenges in detail.

3.1 The Core Role of Sensor Technology

Sensor technology is fundamental for haircutting robots to achieve high precision and intelligent operation. Modern haircutting robots integrate various sensors, such as obstacle avoidance sensors, distance sensors, positioning sensors, and stereo

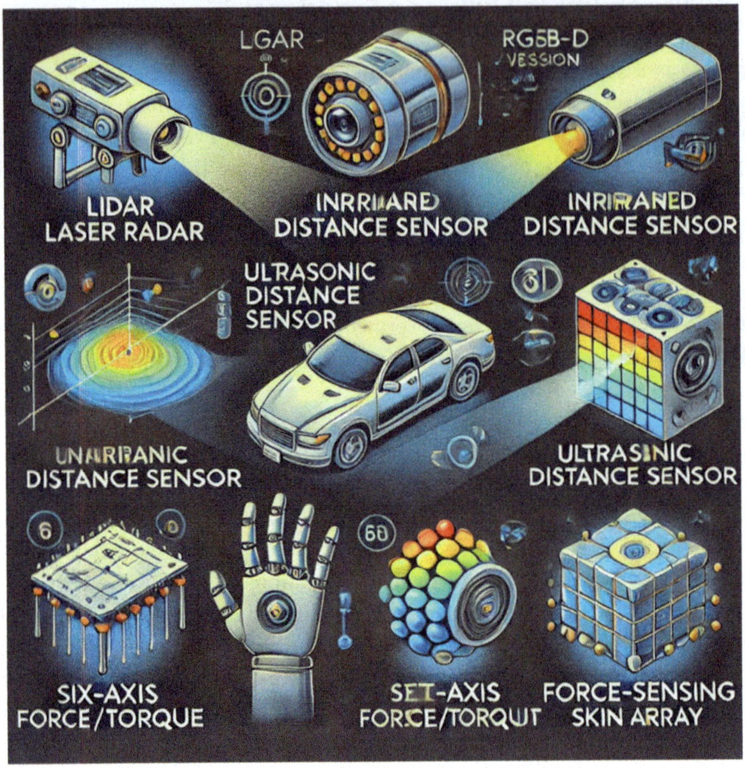

Fig. 3 Advanced sensing technology is a core component of automation and robotics

vision systems, to automate environmental perception and operational control. These sensors enable robots to perceive characteristics of users' hair in real-time, including length, thickness, and density, while effectively avoiding obstacles to ensure user safety [13] (Fig. 3).

Stereo vision technology, as one of the core components, can construct a three-dimensional model of the user's head and analyse hair density and structure in different areas, providing a basis for dynamic adjustments. With this visual perception technology, robots can perform intelligent hairstyle designs and adjust blade angles and cutting force based on real-time detection data, ensuring haircut precision [14]. Sensor technology not only prevents collisions during haircuts but also enhances operational efficiency and haircut quality, minimizing the need for human intervention.

Despite the critical role of sensor technology in haircutting robots, it faces some challenges. For example, accurately identifying and handling external factors in complex environments and effectively integrating data streams from multiple sensors remain technical difficulties. Additionally, the high precision requirements of sensors increase the development costs of equipment [15].

3.2 Communications and Remote Operation Technology

With advancements in remote operation technology, haircutting robots are expected to provide cross-regional haircut services in certain special scenarios. Remote operation technology can help users in remote areas access skilled hairdresser services via internet connectivity, especially in regions with a shortage of human resources, filling the gap between supply and demand [16].

However, remote haircutting robots must overcome challenges posed by communication delays. In real-time operations, even minor delays can result in inaccurate haircuts or harm to users. For instance, when a hairdresser remotely operates a robot, any adjustments to the angle and force of the hair during cutting must respond instantaneously. If delays occur, the robot may fail to respond promptly to the user's head movements or the hairdresser's commands, potentially leading to hairstyle deviations or safety issues [17].

To address this issue, researchers are exploring the application of low-latency communication technologies such as 5G. With its high-speed and low-latency features, 5G networks can significantly reduce delays in remote operations, enhancing the operational accuracy and safety of haircutting robots [18].

Additionally, some haircutting robots can incorporate localized autonomy, allowing them to execute certain basic cutting operations independently without relying on continuous network connectivity, such as adjusting cutting paths or real-time obstacle avoidance (Fig. 4).

3.3 Speed and Safety of Robots

Haircutting robots must not only possess efficient cutting speeds but also ensure safety during the haircutting process. High-speed blades, if not effectively monitored for the surrounding environment and the user's head status, may lead to cuts or erroneous operations. Thus, balancing speed and safety becomes a critical issue in designing high-speed haircutting robots.

The close integration of sensors and control systems is the key to addressing this issue. Obstacle avoidance sensors and distance sensors must have high sensitivity and rapid response capabilities to ensure they can quickly perceive any abnormal changes around the user's head during high-speed operations, such as sudden head movements or irregular shapes [19]. Through a precise feedback mechanism, the robot can adjust its actions in real-time to ensure safety during high-speed operations.

Additionally, haircutting robots must maintain accurate cutting control to ensure that rapid operations do not damage the user's hair or scalp. Force control technology plays a vital role in this process. Robots can perceive the forces applied during haircuts through their force control systems, dynamically adjusting based on hair thickness and hardness. For example, if the robot detects hair that is overly thick or dense, it will automatically reduce cutting speed and force to avoid causing damage to the hair and scalp [20] (Fig. 5).

Fig. 4 Communication delays will significantly impair the performance of remote robotic operations

3.4 Force Control and Feedback Technology

Force control and feedback technologies are crucial for collaborative robots to achieve precise operations, and they are equally applicable in haircutting robots. Traditional industrial robots typically use rigid control systems that struggle to respond to minor changes in soft objects (such as human hair), whereas force control technology allows robots to adjust themselves upon sensing external forces, enabling more flexible and safe operations [21].

The force control system in haircutting robots can detect the forces applied during haircuts in real-time, adjusting the speed and strength of the tools to ensure no damage occurs to fine, brittle, or thick hair. By integrating force feedback technology, robots can sense the resistance of hair and adjust cutting parameters based on real-time feedback data, ensuring smooth and precise haircuts [22].

This technology can also enhance the comfort and safety of haircuts. The force control system not only ensures that the force applied during haircuts is moderate but also prevents the cutting tools from coming into contact with the user's

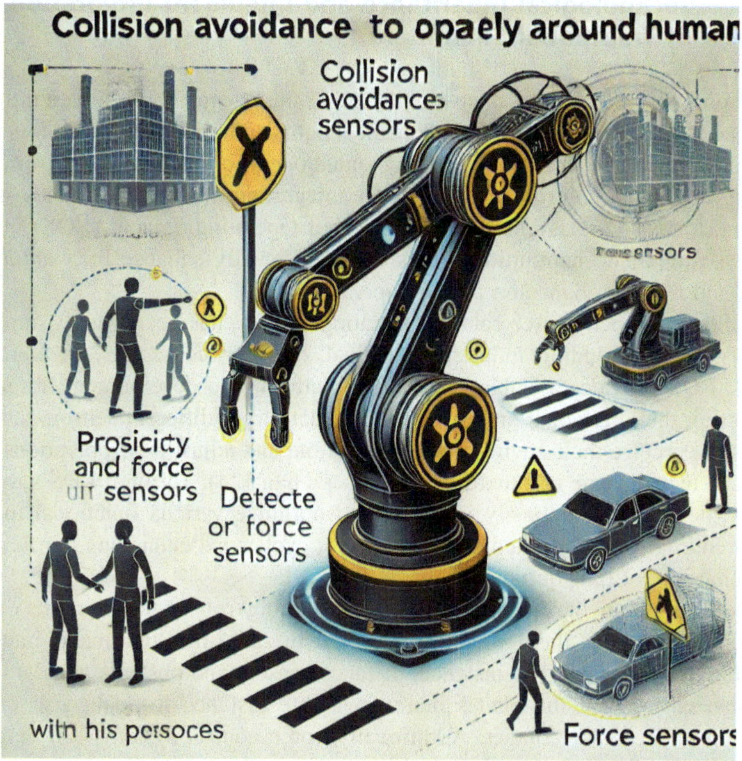

Fig. 5 Advanced intelligent sensing ensures that robots do not pose safety threats to users

scalp, avoiding injury. Furthermore, future developments in force control technology may allow haircutting robots to make more precise adjustments based on users' personalized needs, providing more customized services.

4 Improvements and Integration of Existing Equipment in Robot Design

The design of haircutting robots is not merely a simple automation of traditional haircutting equipment; it incorporates innovative integration with modern robotic technologies. In achieving this goal, haircutting robots inherit and improve upon existing devices like electric clippers, combining intelligent modules and advanced control systems to progressively develop toward efficiency, precision, and safety. The following sections will explore how to improve and integrate existing devices in robot design.

4.1 Technological Inheritance and Intelligent Integration of Traditional Equipment

Traditional electric clippers have already achieved a degree of automation through mechanical design, such as adjusting blade lengths and using electric motors. However, pure mechanization still relies on manual operation and judgment from users, lacking intelligent control capabilities. By integrating intelligent modules such as sensors, force control systems, and the Robot Operating System (ROS) into traditional equipment, haircutting robots can significantly enhance their automation levels and reduce reliance on manual operations [23].

ROS, as an open-source robotic operating system, has been widely applied in various fields, including industrial, medical, and service sectors. In haircutting robots, ROS provides a stable operational platform and development framework, supporting the integration and processing of data from different sensors, enabling robots to perceive their environment in real-time and adjust their operations based on the detected shape and texture of the user's hair [24]. Through ROS's modular design, developers can easily integrate and optimize various functional modules of haircutting robots, shortening development cycles and enhancing the flexibility and reliability of the equipment.

The integration of sensors can help haircutting robots monitor the external environment during the haircutting process in real-time and make intelligent adjustments as needed. For instance, distance sensors and obstacle avoidance sensors can ensure that the blades maintain a safe distance from the user's scalp, avoiding unnecessary contact. Additionally, force control systems can detect the forces applied during haircuts in real-time and automatically adjust the strength of the cut to ensure that hair is not pulled or damaged [25].

4.2 Application of 3D Printing and Path Planning Technology

The layered processing principles of 3D printing technology provide new ideas for precise hairstyle design. Haircutting robots can adopt this principle by performing layer-by-layer trimming to achieve accurate designs. Similar to 3D printing, haircutting robots can gradually trim different areas of hair based on the user's three-dimensional head model, ensuring symmetry and precision in hairstyle design [26].

Path planning technology also plays a significant role in this process. Like the path planning used in CNC machining in industrial manufacturing, haircutting robots need to accurately plan each tool's movement path to ensure that every trim meets the design requirements. With the aid of 3D scanning and path planning algorithms, haircutting robots can automatically generate haircut plans, avoiding repeated trimming or overlooked areas [27]. This automated path planning not only improves haircut efficiency but also reduces human intervention, ensuring that the operational process is more stable and controllable (Fig. 6).

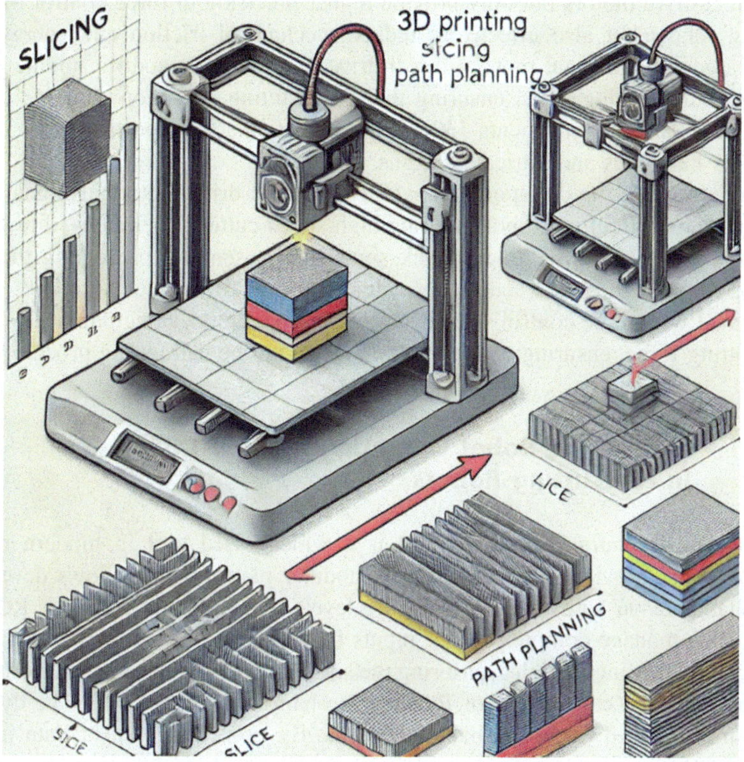

Fig. 6 3D printing technology allows automated clippers to operate along predetermined paths, achieving automatic haircuts, while corresponding software's path planning technology ensures the feasibility of planned paths

Additionally, path planning technology can assist robots in making optimal decisions when handling complex hairstyles. For example, when performing layered trimming, haircutting robots can plan the best trimming path to maintain uniformity across each layer, taking into account variations in hair thickness and density, thereby providing a refined haircutting experience [28].

4.3 Application of Direct-Drive Motor Technology

In robot design, direct-drive motor technology is an essential component of haircutting robots due to its advantages in rapid response and high precision control. Traditional electric clippers rely on transmission devices to regulate tool movement, but this approach often introduces mechanical delays that affect cutting accuracy. Direct-drive motors, on the other hand, directly drive the cutting tools, eliminating intermediate transmission links and significantly enhancing the equipment's response speed [29].

Direct-drive motors not only provide higher precision in force control for haircutting robots but also effectively reduce mechanical friction and energy loss. Compared to traditional motors, direct-drive motors better control minute movements of the cutting tools, ensuring that each cutting operation achieves precise length and angle adjustments [30]. This is particularly important for handling complex hairstyles and intricate designs.

Moreover, the rapid response capability of direct-drive motors also guarantees the safety of haircutting robots. During high-speed cutting operations, direct-drive motors can react swiftly to feedback signals from sensors, preventing the tools from contacting the user's scalp, thus reducing the risk of accidental injuries. When combined with force control systems, direct-drive motors can automatically adjust the cutting force, ensuring user comfort throughout the haircutting process [31].

4.4 Role of the Robot Operating System (ROS) in Haircutting Robots

The Robot Operating System (ROS) is a widely used tool in modern robotic development, providing a standardized, modular platform that allows developers to collaborate on hardware and software levels. For haircutting robots, ROS not only helps manage complex sensor inputs but also coordinates operations among multiple processing modules, ensuring the smooth functioning of the entire system.

The real-time capabilities of ROS enable haircutting robots to make dynamic adjustments based on users' hairstyle needs. By processing sensor data in real-time, ROS can adjust cutting strategies according to hair thickness, length, and shape. Additionally, ROS supports the integration of remote operations and automated functions, allowing developers to quickly implement remote haircutting capabilities and even control haircutting robots via cloud-based systems [32].

In the future, ROS may integrate more artificial intelligence algorithms, enabling haircutting robots to learn users' haircutting habits and preferences more intelligently, automatically generating personalized hairstyle design proposals. By combining with machine learning, haircutting robots can suggest new hairstyles based on users' haircut histories and facial structures, further enhancing the haircutting experience.

5 Future Prospects: Evolution from Automation to Full Intelligence

The development of haircutting robots is moving towards full automation and intelligence. In the future, with advancements in artificial intelligence, machine learning, and robotic technologies, haircutting robots will not merely be automated devices but intelligent systems that can self-adjust and adapt to different user needs. Their application scenarios will extend from homes to professional salons and nursing homes. At the same time, innovations in smart devices and sustainable development concepts will further promote the popularization and optimization of haircutting robots.

5.1 Expansion of Application Scenarios for Fully Automated Haircutting Robots

In the future, haircutting robots will not only be a choice for household users but will also play important roles in various professional settings. As technology matures, the costs of automated haircutting equipment will further decrease, and performance will become more stable, allowing diverse services for different user groups.

In home settings, haircutting robots will greatly simplify personal haircutting processes, especially for those unwilling to frequently visit salons. Household haircutting robots will offer various hairstyle presents, allowing users to choose their preferred style, which the robot will then automatically execute based on their hair type and head shape. Such devices will reduce users' time and financial costs, providing a convenient and quick home haircutting experience [33].

In professional salons, haircutting robots can assist hairdressers in handling numerous basic cutting tasks. For example, robots can perform layered haircuts with precise cutting algorithms, ensuring consistency and accuracy, allowing hairdressers to focus on more complex hairstyle designs and personalized services. This human–robot collaboration model will improve salon efficiency, reduce the probability of human errors, and enhance customer satisfaction [34].

In nursing homes and hospitals, haircutting robots will provide more convenient services for the elderly and those with mobility issues. Such populations often struggle to visit salons regularly, and nursing homes and hospitals can use haircutting robots to provide regular personal care. These automated devices can adjust cutting force and speed based on users' physical conditions, ensuring safety and comfort [35].

By expanding application scenarios, haircutting robots will not only meet individual user needs but will also play significant roles in public service and professional settings, becoming valuable assistants in daily life and work.

5.2 Innovative Directions for Smart Haircutting Devices

With advancements in artificial intelligence and machine learning technologies, future haircutting robots will not just execute present tasks but will also possess the ability to learn and optimize autonomously, further enhancing their intelligent and personalized service levels.

The application of artificial intelligence algorithms will enable haircutting robots to analyse vast amounts of user data, learning and optimizing the haircutting process. For instance, robots can record and analyse users' hair characteristics, hairstyle preferences, and haircut histories to generate personalized hairstyle proposals. Deep learning algorithms can assist haircutting robots in recognizing users' facial and head shapes, recommending the most suitable hairstyles based on big data analysis [11]. This personalized hairstyle design will improve user experience, saving time and enhancing haircut results.

The application of machine learning will allow haircutting robots to continuously optimize cutting strategies with each use. Through feedback mechanisms, robots can adjust cutting force, speed, and angles based on user feedback, progressively improving service quality. Future haircutting robots may even actively recommend new hairstyles to users by continuously learning the relationships between various hairstyles and users' facial structures [36].

This intelligent innovation direction not only enhances the intelligence level of haircutting robots but also provides users with more personalized services, breaking through the limitations of traditional haircutting equipment.

5.3 Sustainable Development and Ecological Design

In the context of growing global attention to environmental protection and sustainable development, the design and manufacturing of haircutting robots will gradually move towards eco-friendly and high-efficiency directions. Future haircutting equipment will not only see functional upgrades but will also utilize materials and energy consumption that are more environmentally friendly, aligning with sustainable development needs.

The use of environmentally friendly materials will become a significant direction in the design of future haircutting robots. The shells, components, and consumables of haircutting robots can utilize recyclable or biodegradable materials, thereby reducing negative impacts on the environment. For example, the plastics used in external devices can be replaced with biodegradable materials, and the consumables (such as blades) inside the robots can be designed to minimize environmental harm through technological innovations [37].

Energy efficiency optimization and low-power designs will be another key area for future haircutting equipment. With advancements in energy technology, haircutting robots will gradually adopt more efficient motors and control systems, ensuring performance while reducing power consumption. In long-term usage scenarios (such as nursing homes or households), low-power devices will reduce energy consumption and operating costs [38]. Additionally, future haircutting robots may integrate solar energy or other renewable energy technologies to provide green energy solutions for the equipment.

Intelligent waste management systems will also be an innovative aspect of future haircutting robots. With technological advancements, haircutting robots could be equipped with built-in intelligent waste disposal systems to automatically collect and process cut hair, minimizing environmental pollution. In the future, there may even be specialized technologies to repurpose this hair for reuse, such as using it as agricultural fertilizer or producing eco-friendly materials [39].

By incorporating sustainable development concepts, haircutting robots will not only upgrade functionally but will also contribute to environmental protection, further advancing smart devices towards greener and more environmentally friendly directions.

References

1. Hastie, T., Tibshirani, R., & Friedman, J. (2009) *The elements of statistical learning: Data mining*, Inference, and Prediction, 2nd ed. Springer
2. Kober, J., Bagnell, J. A., & Peters, J. (2013). Reinforcement learning in robotics: A survey. *The International Journal of Robotics Research, 32*(11), 1238–1274.
3. D'Andrea, R. (2014). Guest editorial can drones deliver? *IEEE Transactions on Automation Science and Engineering, 11*(3), 647–648.
4. Tjeng, V., Xiao, K., & Tedrake, R. (2017). Evaluating robustness of neural networks with mixed integer programming. *arXiv preprint* arXiv:1711.07356.
5. Ott, C., Albu-Schäffer, A., Kugi, A., & Hirzinger, G. (2008). On the passivity-based impedance control of flexible joint robots. *IEEE Transactions on Robotics, 24*(2), 416–429.
6. Derpanis, K. G. (2010). Overview of the RANSAC algorithm. *Image Rochester NY, 4*(1), 2–3.
7. Sutton, R., & Barto, A. G. (2018). *Reinforcement learning: An introduction*, 2nd ed. MIT Press.
8. Corke, P. (2017). *Robotics, vision and control: Fundamental algorithms in MATLAB*, 2nd ed. Springer.
9. Thrun, S., Burgard, W., & Fox, D. (2005). *Probabilistic robotics*, MIT Press.
10. Peters, J., Vijayakumar, S., & Schaal, S. (2003). Reinforcement learning for humanoid robotics. In *Proceedings of the 2003 IEEE-RAS International Conference on Humanoid Robots*, Karlsruhe, Germany, pp. 1–20.
11. Silver, D., Huang, A., Maddison, C. J., et al. (2016). Mastering the game of Go with deep neural networks and tree search. *Nature, 529*(7587), 484–489.
12. Deisenroth, M. P., Neumann, G., & Peters, J. (2013). *A survey on policy search for robotics, foundations and trends® in robotics*.
13. Lynch, K. M., & Park, F. C. (2017). *Modern robotics: mechanics, planning, and control*, Cambridge University Press.
14. Kroemer, O., Niekum, S., & Konidaris, G. (2021). A review of robot learning for manipulation: Challenges, representations, and algorithms. *Journal of machine learning research, 22*(30), 1–82.
15. Nguyen-Tuong, D., & Peters, J. (2011). Model learning for robot control: A survey. *Cognitive processing, 12*, 319–340.
16. Schwartz, M. (2019). *5G Mobile networks: A systems approach*, 1st ed. Springer.
17. Siciliano, B., & Khatib, O. (2016). *Springer handbook of robotics*, 2nd ed. Springer.
18. Chataut, R., & Akl, R. (2020). Massive MIMO systems for 5G and beyond networks—overview, recent trends, challenges, and future research direction. *Sensors, 20*(10), 2753.
19. Raibert, M. H., Blankespoor, K., Nelson, G. et al. (2008). BigDog, the rough-terrain quadruped robot. In *Proceedings of the 17th International Federation of Automatic Control World Congress*, Seoul, Korea, pp. 10822–10825.
20. Albu-Schäffer, A., Ott, C., Frese, U., & Hirzinger, G. (2003) Cartesian impedance control of redundant robots: Recent results with the DLR-light-weight-arms. In *Proceedings of the 2003 IEEE International Conference on Robotics and Automation (ICRA)*, Taipei, Taiwan, pp. 3704–3709.
21. Peters, J., Kober, J., Dillmann, R., & Schaal, D. (2008). Reinforcement learning for humanoid robotics. *IEEE Transactions on Robotics, 24*(2), 374–385.
22. Hauser, K., Bretl, T., Harada, K., & Latombe, J. C. (2008). Motion planning for legged robots on varied terrain. *The International Journal of Robotics Research, 27*(11–12), 1325–1349.

23. Quigley, A., Smart, C., & Stoyanov, T. (2017). *ROS Robotics by example*, 2nd ed. Packt Publishing.
24. Diankov, R., & Kuffner, J. (2010). OpenRAVE: A planning architecture for autonomous robotics. *IEEE Transactions on Automation Science and Engineering, 7*(2), 146–154.
25. Nardi, D., Roberts, J., Veloso, M., & Fletcher, L. (2016). Robotics competitions and challenges. *Springer handbook of robotics*, pp. 1759–1788.
26. Gul, J. Z., Sajid, M., Rehman, M. M., Siddiqui, G. U., Shah, I., Kim, K. H., Lee, J. W., & Choi, K. H. (2018). 3D printing for soft robotics–a review. *Science and Technology of Advanced Materials., 19*(1), 243–262.
27. Zhang, Z., Wu, L., Zhang, W., Peng, T., & Zheng, J. (2021). Energy-efficient path planning for a single-load automated guided vehicle in a manufacturing workshop. *Computers & Industrial Engineering, 158*, 107397.
28. Fahimi, F., Nataraj, C., & Ashrafiuon, H. (2009). Real-time obstacle avoidance for multiple mobile robots. *Robotica, 27*(2), 189–198.
29. Tan, K. K., Lee, T. H., & Huang, S. (2007). *Precision motion control: design and implementation.* Springer Science & Business Media.
30. Nasr, A., Gu, C., Wang, X., Buticchi, G., Bozhko, S., & Gerada, C. (2021). Torque-performance improvement for direct torque-controlled PMSM drives based on duty-ratio regulation. *IEEE Transactions on Power Electronics, 37*(1), 749–760.
31. Perrusquía, A. (2022). Robust state/output feedback linearization of direct drive robot manipulators: A controllability and observability analysis. *European Journal of Control, 64*, 100612.
32. Spong, M. W., Hutchinson, S., & Vidyasagar, M. (2020). *Robot modeling and control.* John Wiley & Sons.
33. Gonzalez-Aguirre, J. A, Osorio-Oliveros, R., Rodriguez-Hernandez, K. L., Lizárraga-Iturralde, J., Morales Menendez, R., Ramirez-Mendoza, R. A., Ramirez-Moreno, M. A., de Jesus Lozoya-Santos, J. (2021). Service robots: Trends and technology Applied Sciences, 11. 22: 10702
34. Rodriguez-Guerra, D., Sorrosal, G., Cabanes, I., & Calleja, C. (2021). Human-robot interaction review: Challenges and solutions for modern industrial environments. *IEEE Access, 9*, 108557–108578.
35. Apraiz, A., Lasa, G., & Mazmela, M. (2023). Evaluation of user experience in human–robot interaction: A systematic literature review. *International Journal of Social Robotics, 15*(2), 187–210.
36. Kober, J., & Peters, J. (2011). Policy search for motor primitives in robotics. *Machine Learning, 84*(1–2), 171–203.
37. Sartal, A., Bellas, R., Mejías, A. M., & García-Collado, A. (2020). The sustainable manufacturing concept, evolution and opportunities within Industry 4.0: A literature review. *Advances in Mechanical Engineering, 12*(5), 1687814020925232.
38. Kuehl, A., Kreitlein, S., Hoeft, A., & Franke, J. (2018, February). Energy efficient processes of electric motor manufacturing. In *2018 IEEE Texas Power and Energy Conference (TPEC)* (pp. 1–5). IEEE.
39. Gupta, A. (2014). Human hair "waste" and its utilization: Gaps and possibilities. *Journal of waste management, 2014*(1), 498018.

Mechanical Design and End Effectors

9

1 Introduction

Mechanical design plays a crucial role in the functionality of haircutting robots. One of the core functions of haircutting robots is to accurately and safely trim hair of varying lengths and thicknesses while adapting flexibly to the contours of the user's head. In this process, the design of the mechanical structure directly determines whether the robot can perform efficiently, accurately, and gently.

In haircutting robots, mechanical design involves not only the realization of machine movement but also how to precisely position the hair clippers (end effectors) and execute cutting actions at the appropriate angles. This requires designers to simplify control systems as much as possible while ensuring the freedom of motion in the system, achieving efficient and stable operational performance. The quality of mechanical design directly affects the overall performance of haircutting robots, including the smoothness of movement over complex head surfaces, cutting accuracy, and operational safety. Key challenges faced by haircutting robots include the following:

1. Precision Requirements: During the haircutting process, the robot must not only reach specific positions but also operate with millimeter precision. Particularly when handling different lengths or layers of hairstyles, the robot must precisely control the position and angle of the hair clippers to ensure even trimming.
2. Safety: The head is a vital part of the human body, and haircutting robots must ensure that they do not harm the user during operation. Therefore, the robot needs reliable collision detection and force control capabilities, able to stop quickly in emergencies or maintain appropriate cutting pressure to avoid skin injuries.
3. Adaptability: Each person's head shape varies, so haircutting robots need to possess high adaptability, adjusting based on different head shapes. This involves not only position control but also precise control of the posture of the

© The Author(s), under exclusive license to Springer Nature Switzerland AG 2025

S. Li, *Haircutting Robots*, https://doi.org/10.1007/978-3-031-84026-5_9

hair clippers to ensure close contact with the head surface during the cutting process.

Another important aspect of mechanical design is how to reduce costs while improving maintainability. If the robot is complex and components are expensive, it increases manufacturing costs and poses challenges for subsequent maintenance. Therefore, selecting suitable materials, optimizing structures, and using standardized components are effective ways to lower costs.

Related studies in the literature have presented similar challenges and solutions regarding the mechanical design of robot end effectors. For example, Abdel-Malek et al. emphasized the importance of multi-degree-of-freedom and high-rigidity structures for precision and operational flexibility in the design of parallel robots [1]. Additionally, Sun et al. explored the safety design of industrial robots in complex environments, suggesting redundancy design and sensor fusion to enhance safety performance [2]. The theories and practical experiences from these studies are relevant for the design of haircutting robots.

2 Mechanical Structure of Haircutting Robots

The core design concept of haircutting robots is to combine hair clippers with a freely movable platform to achieve precise trimming of different areas of the head. This platform needs to not only move in space but also change posture to ensure that the hair clippers can process hair from different angles. Through this structural design, haircutting robots can perform complex three-dimensional movements over the head surface, enabling trimming at any position and angle (Fig. 1).

2.1 Key Elements of the Structure

In the mechanical structure design of haircutting robots, there are two main key elements: position control and posture control.

Position Control (*XYZ* Space): Haircutting robots need to move precisely in three-dimensional space to reach any position on the head. Typical displacement control methods include the Cartesian coordinate system (e.g., structures used in 3D printers and CNC machines), where linear motors or screw-driven motors convert rotational motion into linear motion. This linear motion can be achieved through three independent axes, namely the X, Y, and Z axes, to cover the entire trimming range of the head [3].

Posture Control (Pitch, Yaw, and Roll around the Head): In addition to position control, haircutting robots must be able to adjust the posture of the hair clippers to maintain the appropriate angle of the blades relative to the hair surface. This includes controlling three rotational degrees of freedom: pitch, yaw, and roll. Pitch control determines the forward and backward tilt angle of the hair clippers, yaw control adjusts the left and right rotation angle, and roll controls the rotation of

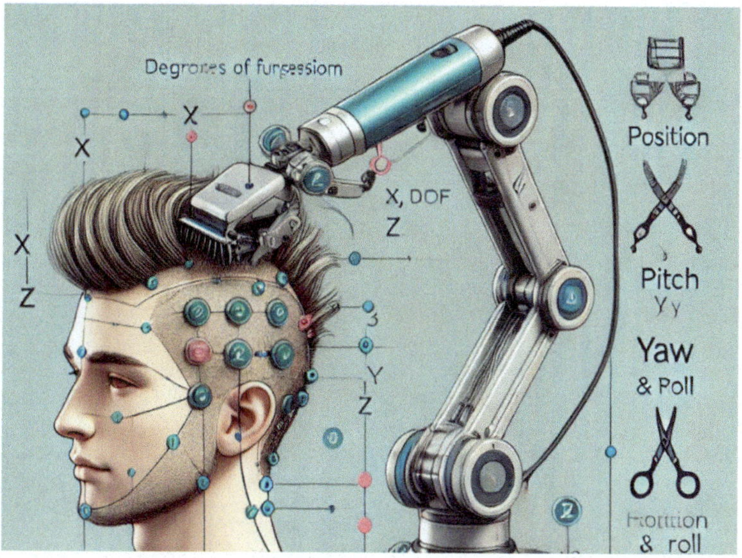

Fig. 1 Schematic diagram of the overall structure of the haircutting robot

the clippers around their own axis. These angle controls ensure that the hair clippers can conform closely to the contours of the head, allowing for precise cutting operations [4].

2.2 Overview of Robot Kinematics

The kinematics of haircutting robots can be divided into position control and posture control. Typically, position control is achieved through three linear axes, while posture control is accomplished through rotational axes. This design is very common in robotics and has multiple implementation methods:

Cartesian robots use an *XYZ* three-axis system to control the displacement of each axis through linear motion. The advantage of this design is its simplicity and ease of control, making it suitable for large-scale movements over the head's surface. However, its drawback is the limited flexibility in posture control, often requiring additional rotational axes to supplement posture adjustments.

Articulated robots control position and posture through the rotation of multiple joints. A typical articulated robotic arm can handle both position and posture simultaneously, providing high flexibility. By adjusting the angle of each joint, the robot can perform more complex movements around the head, adapting to different head shapes and hairstyle requirements [5] (Fig. 2).

Parallel robots (e.g., Delta robots) achieve high-speed and high-precision motion by simultaneously driving the end effector through multiple linkages. This structure typically has high rigidity and fast response characteristics, but its

Fig. 2 Industrial articulated robotic arm

workspace is smaller, requiring specific optimization for the shape of the head when applied to haircutting robots [6].

By combining position control and posture control, haircutting robots can perform cutting operations at any position on the head while adjusting the angle of the hair clippers to ensure cutting precision and effectiveness. Additionally, by utilizing redundant degrees of freedom (e.g., adding extra rotational axes), the robot's movement flexibility over complex head contours can be further enhanced, preventing collisions.

2.3 Challenges in Practical Application.

Although the above mechanical structures theoretically address the movement needs of haircutting robots effectively, several challenges still exist in practical applications. For instance, improving movement speed while ensuring high precision and avoiding collisions around the head are issues that require further exploration. By optimizing mechanical design, especially in terms of redundancy in posture control, haircutting robots may achieve smooth operations while reducing degrees of freedom.

3 Position Control Mechanism

In the design of haircutting robots, position control is critical for determining whether tasks can be executed precisely. To ensure that the robot can perform fine trimming in different areas of the head, a flexible and reliable displacement control system must be designed. There are several existing position control methods, each with its unique characteristics, among which Cartesian coordinate systems, rotational coordinate systems, and parallel mechanisms are the most common.

First, the Cartesian coordinate system is a classic motion control method widely used in 3D printers and CNC machines. In this system, movement is achieved through three independently controlled axes corresponding to the X, Y, and Z directions. The motion of each axis can be driven by motors, commonly implemented using screw drives that convert rotational motion into linear movement, allowing precise control of the end effector's position in space. The advantages of this design lie in its simplicity and relatively easy control algorithms, making it highly suitable for applications that require high precision and controllable motion paths. Additionally, to further enhance speed and accuracy, linear motors can be used for direct drive, eliminating intermediate mechanical transmission. This method can provide faster response times and higher precision, thus being widely adopted in some high-end devices [7].

However, the limitations of the Cartesian coordinate system are also quite evident. Although it can provide high-precision spatial displacement control, it often struggles with complex posture adjustments. Particularly when the haircutting robot needs to perform multi-angle, complex movements on the head's surface, relying solely on the Cartesian coordinate system may not meet all demands. To address this issue, additional rotational axes are usually needed for posture control.

In contrast, rotational coordinate systems offer significant advantages in flexibility. They achieve complex three-dimensional movements through the design of articulated robotic arms, utilizing the collaborative action of multiple rotational joints. Each joint's rotation angle is precisely controlled by motors, allowing the robotic arm to easily maneuver around obstacles on the head to reach difficult trimming areas. For instance, when trimming the sides or back of the head, the rotational coordinate system can easily adjust the position of the hair clippers through multi-joint movements. This design not only enhances the flexibility of haircutting robots but also allows for angle adjustments of the clippers, ensuring that the blades closely conform to the scalp during cutting [8].

Despite the high flexibility of rotational coordinate systems, their complex joint structures make control algorithms relatively intricate, and additional joints may lead to decreased load capacity and precision. Therefore, during the design process, it is essential to weigh flexibility against system complexity based on specific application requirements.

Parallel mechanisms represent another common mechanical structure, with Delta robots being a typical example. Unlike Cartesian and rotational coordinate systems, parallel mechanisms control the end effector's position through multiple linkages. The advantages of this design include high rigidity and fast response, making it particularly suitable for applications requiring high-speed, high-precision movements. In haircutting robots, parallel mechanisms can be used to quickly complete trimming tasks over large areas of the head. For example, the Delta robot structure allows for rapid movement of the hair clippers through the coordinated length changes of multiple linkages while maintaining high motion precision. This high-speed capability provides significant advantages in tasks requiring extensive repetitive movements [9].

However, the downside of parallel mechanisms is that they have relatively small workspaces, making it difficult to cover the entire head's trimming area. Therefore, parallel mechanisms are generally suitable for specific local tasks rather than comprehensive haircutting work. Designers need to pay particular attention to the workspace limitations when choosing parallel mechanisms and plan their applications within the robotic system accordingly.

In summary, different control methods in the design of haircutting robots each have their advantages and disadvantages. The Cartesian coordinate system is suitable for large-scale precise control but has limitations in posture adjustments; the rotational coordinate system offers high flexibility but comes with increased control complexity; parallel mechanisms excel in high-speed, high-precision operations but have smaller workspaces. Reasonable selection and combination of these control methods based on specific application needs and design objectives are crucial for realizing efficient and precise haircutting robots.

4 Degrees of Freedom (DoF) Requirements

The design of haircutting robots must consider spatial movement and posture control, and the number of degrees of freedom (DoF) directly impacts the robot's operational flexibility and accuracy. Controlling the position and posture of the clippers to ensure precise trimming is particularly important during the haircutting process. Next, we will explore the DoF requirements for position and posture control in haircutting robots and analyze how introducing redundant DoF can enhance operational flexibility and safety.

4.1 Degrees of Freedom in Position Control

Typically, three degrees of freedom (XYZ) are sufficient to meet spatial position control requirements. These three degrees correspond to the robot's linear movements in three-dimensional space, similar to achieving coverage of different areas of the head through displacements along the X, Y, and Z axes. The core of position control lies in the ability to move the hair clippers precisely to reach various areas of the head for trimming. Whether using a Cartesian coordinate system or an articulated robot, a design with three degrees of freedom ensures that the hair clippers can move along the head's surface to meet basic trimming needs [10].

However, position control alone is not enough. Controlling the posture of the hair clippers is equally important, as the blades must contact the hair at specific angles during cutting, requiring the robot not only to reach the target position but also to adjust the angle of the hair clippers.

4.2 Degrees of Freedom in Posture Control

To control the posture of the hair clippers relative to the head's contour, typically an additional two degrees of freedom are needed. These two degrees allow the hair clippers to be adjusted at different angles, ensuring that the blades maintain contact with the scalp throughout the trimming process. These two degrees correspond to the pitch and yaw of the clippers, enabling them to adapt to various curves on the head's surface.

If we consider that the robot needs to achieve complete movement and posture adjustments in space, a six-degree-of-freedom design is required. Three degrees are used for position control, while the other three are for posture control (including pitch, yaw, and roll). This design provides sufficient flexibility for the robot to perform precise trimming at any position and angle on the head [5].

4.3 Introduction of Redundant Degrees of Freedom

In some cases, redundant degrees of freedom may be necessary. Although a six-degree-of-freedom design theoretically covers most haircutting needs, practical operations can encounter collision issues due to the complexity of head structures and potential obstacles during haircuts (e.g., ears, hairlines). To avoid this, designers can introduce additional degrees of freedom, allowing the robot to maneuver around these obstacles and successfully complete tasks.

For example, adding an extra joint or rotational axis can provide the robot with more path options, preventing collisions between the robotic arm and the head during haircuts. This design of redundant degrees of freedom not only enhances the robot's operational flexibility but also prevents interruptions in trimming due to limited movement paths [11].

4.4 Managing Degrees of Freedom in Optimization Design

In the design of degrees of freedom, how to maintain the robot's flexibility while controlling costs and complexity is an important design consideration. Although increasing degrees of freedom can improve flexibility in confined spaces, more degrees also mean more complex control systems and higher manufacturing costs. Therefore, designers can optimize movement paths and mechanism designs to achieve similar operational effects with fewer degrees of freedom.

For instance, a six-degree-of-freedom design can avoid collisions when trimming specific areas of the head by carefully planning movement paths. Such optimized designs can reduce degrees of freedom while ensuring robot functionality, lowering production costs and system complexity. Additionally, utilizing advanced path planning algorithms can enhance the robot's flexibility under limited degrees of freedom through intelligent control [12].

In summary, the choice of degrees of freedom in the design of haircutting robots is not only related to functional realization but also affects the overall system efficiency and cost. Reasonably configuring the degrees of freedom required for position and posture control, and introducing redundant degrees when necessary, are key to ensuring the efficient and safe operation of haircutting robots.

5 Mechanical Design Principles for Haircutting Robots

When designing haircutting robots, the precision, flexibility, and spatial efficiency of the mechanical structure are critical factors. Particularly for a system that needs to operate efficiently in the confined space of a head, achieving a balance between compact design and flexibility is one of the main challenges faced by designers. To this end, we can draw on the design experiences of similar mechanical systems, such as five-axis CNC machines, while combining compact and scalable design strategies to ensure that haircutting robots can perform complex motion tasks in small spaces.

5.1 Comparative Analysis of Five-Axis CNC Machines

Five-axis CNC machines provide a classic example of controlling complex movements. This system typically uses three linear axes (X, Y, Z) and two rotational axes (A, B) to achieve complete control over tool position and posture. In the design of haircutting robots, similar multi-axis systems can provide sufficient degrees of freedom to enable the hair clippers to operate precisely at various positions on the head while adjusting the blade angle to accommodate different hairstyle needs.

The design philosophy of five-axis CNC systems emphasizes the combination of position control and posture control. The movement of three linear axes ensures that the hair clippers can reach any position on the head, while the two rotational axes are used to adjust the clippers' angle, ensuring the blades maintain the optimal cutting angle against the head's surface. This combination provides greater operational space and flexibility for haircutting robots, enabling them to tackle complex head geometries [13] (Fig. 3).

Fig. 3 Five-axis CNC machining for engraving

In the design of a haircutting robot, designers need to consider how to balance compactness, precision, and workspace efficiency. One of the design principles of five-axis CNC machines is to optimize the arrangement of axes and movement paths to minimize the space required for tool movement, thereby enhancing system compactness. This concept is equally applicable to haircutting robots, especially when operating within the limited space around the head. How to achieve the maximum range of motion with the minimal mechanical structure becomes a key challenge.

5.2 Strategies for Compact and Expandable Design

In the design of haircutting robots, both compact and expandable designs have their advantages and drawbacks. Compact designs focus on minimizing the volume and complexity of mechanical structures, allowing the robot to operate efficiently in the constrained space around the head. The advantage of such designs is that they can achieve the necessary degrees of freedom without occupying too much space. However, compact designs may limit the robot's operational flexibility, especially when significant posture adjustments are required. A compact structure might not provide sufficient rotational space [14].

One core strategy of compact design is optimizing the layout of the axes. For example, by integrating multiple movement axes into a single joint, the overall length and volume of the robotic arm can be reduced. Additionally, modular design is a common compact strategy, which uses standardized mechanical modules to meet complex movement needs while maintaining structural simplicity.

In contrast, expandable designs prioritize providing a larger workspace and greater flexibility. In this type of design, the haircutting robot can expand its range by adding more degrees of freedom or using a longer robotic arm. This ensures that the robot can cover all areas of the head and navigate around obstacles or handle complex hairstyles with greater adaptability [15].

However, expandable designs often come with increased structural complexity. More movement axes mean more complicated control systems, which may also increase manufacturing costs and maintenance difficulties. Therefore, when choosing between compact and expandable designs, designers must balance operational flexibility with system complexity based on actual application requirements.

5.3 Trade-Offs and Optimization in Design

In the mechanical design of haircutting robots, the trade-off between compactness and expandability ultimately depends on the robot's task requirements. For example, if the robot's primary task is to perform simple cutting operations with low posture control requirements, a compact design may be more suitable. However, if the robot needs to handle complex hairstyle designs and diverse head shapes, an expandable design will provide greater flexibility and adaptability.

Additionally, the introduction of intelligent control systems and path-planning algorithms in modern mechanical design can help optimize design, reducing the dependency on physical degrees of freedom. For instance, by using path-planning algorithms, the robot can achieve complex operations with fewer degrees of freedom by optimizing movement paths. This not only reduces the complexity of the mechanical structure but also improves system efficiency and responsiveness [16].

In conclusion, the design of haircutting robots must comprehensively consider compactness, flexibility, and system complexity. By carefully selecting mechanical structures and control schemes, designers can meet the practical needs of automated haircuts.

6 Mechanical Design of the Haircutting Clippers

As the core end effector of a haircutting robot, the mechanical design of the clippers determines the precision, comfort, and safety of the haircutting process. During the haircut, the clippers must precisely cut hair of different lengths while avoiding damage to the scalp. Therefore, the mechanical structure of the clippers must not only provide efficient cutting functionality but also offer a flexible adjustment mechanism. Below, we will discuss the adjustable cutting mechanism and the mechanical realization of the cutting and sliding actions.

6.1 Adjustable Cutting Mechanism

One of the key functions of the haircutting clippers is adjusting the distance between the blades and the comb based on the desired hair length. Traditional manual clippers rely on the barber to make these adjustments, while the haircutting robot requires an automated mechanism to quickly adjust based on different hairstyle needs.

The core of the automatic adjustment mechanism lies in the integration of sensors and actuators, allowing the clippers to adjust the cutting distance based on real-time feedback. For example, a miniature stepper motor can be installed between the blades and the comb, and by precisely controlling the motor's steps, the gap between the blade and the comb can be changed. This design ensures that the haircutting robot can quickly switch the cutting length of the blade for different haircutting tasks, meeting the needs for various hair lengths [17].

The combination of sensors and actuators is an essential part of this design. Using optical or ultrasonic sensors, the clippers can detect the current hair length and automatically adjust the distance between the blades and the comb through a feedback system. This adaptive adjustment mechanism allows the haircutting robot to provide more efficient and precise haircutting services based on the personalized needs of each user.

6.2 Mechanical Realization of Cutting and Sliding Actions

The mechanical realization of the clippers involves both the cutting and sliding actions. During the haircut, the clippers must perform the cutting action smoothly and precisely, avoiding pulling the hair or causing unnecessary resistance.

The cutting action relies on a high-precision motor-driven system, typically a reciprocating blade mechanism powered by a motor. This mechanism operates similarly to traditional electric clippers, cutting hair through the high-speed reciprocating movement of the blades. To ensure the smoothness of the cutting process, the blade design needs to take the following points into consideration:

1. Sharpness of the blades: The blades must be sharp enough to cut the hair cleanly in one pass without pulling.
2. Cutting stability: By optimizing the blade material and geometry, vibration and noise during cutting can be reduced, improving user comfort [18].

The sliding action mainly involves the clippers gliding over the scalp. By designing a suitable bottom structure for the clippers, smooth movement can be ensured, preventing the clippers from getting stuck or irritating the scalp. The smoothness of the sliding action is also related to the geometry and arrangement of the blades, all of which must be carefully considered during design to ensure the robot can complete the haircut smoothly.

6.3 Comfort and User Safety in Haircutting

User safety and comfort must always be the top priority when designing the mechanical structure of the haircutting clippers. During the haircut, the clippers must not only perform efficiently but also ensure that no harm is caused to the user's scalp. Various methods can be employed in the design to enhance the comfort of the haircut:

Smooth cutting action: By optimizing the geometry and cutting path of the blades, smooth blade movement during the haircut can be ensured, reducing hair pulling.

Optimizing the contact surface with the scalp: By combining flexible materials and force-control algorithms, the clippers can move over the scalp without applying excessive pressure or friction, preventing discomfort or injury to the scalp.

Through the above designs and improvements, the clippers not only achieve efficient and precise haircutting but also ensure the safety and comfort of the user, thus providing the haircutting robot with a stable and reliable end effector.

7 Considerations in Mechanical Manufacturing

In the design and manufacturing process of haircutting robots, the quality of the mechanical structure directly determines the robot's performance and durability. Material selection, manufacturing precision, and production cost control are three critical factors that together influence whether the robot can be competitive in the market and provide efficient, stable services. Below, we will explore the key considerations in the manufacturing of haircutting robots, focusing on material selection, manufacturing precision, and cost-effectiveness.

7.1 Material Selection

Material selection is crucial in the manufacturing of haircutting robots. The robot needs to use lightweight, durable, and corrosion-resistant materials to ensure stable long-term operation while not causing discomfort to the user during haircuts. For a haircutting robot, the materials not only determine the strength and rigidity of the mechanical components but also directly affect the overall weight, manufacturing costs, and the robot's lifespan.

First, lightweight materials can effectively reduce the total weight of the robot, improving operational flexibility and speed. For example, aluminum alloys are a common choice. They offer high strength and rigidity while being lightweight and easy to machine into complex mechanical parts. Aluminum alloys also have excellent corrosion resistance, which is important for dealing with the dampness and wear typical of a haircutting environment [19].

In addition, plastic composites have been widely used in some non-load-bearing components. These materials are not only cost-effective but can also be processed into complex shapes through methods such as injection molding. Modern engineering plastic composites, such as carbon fiber-reinforced plastics, offer excellent stiffness and strength while maintaining a lightweight nature. This makes them ideal for use in components like the outer shell and protective covers of haircutting robots [20].

For components that come into direct contact with hair during the haircutting process, the use of corrosion-resistant materials such as stainless steel is necessary. Stainless steel is not only durable but also resists rust and corrosion over long-term use, ensuring the sharpness and precision of the hair clipper blades and other critical components [21].

7.2 Manufacturing Precision

Many key components in haircutting robots require extremely high manufacturing precision, especially in mechanical parts related to position control and cutting precision. The robot must be able to achieve smooth and precise movements over the complex curves of the head, which means that the tolerance of each moving part must be strictly controlled.

Precision manufacturing technologies such as CNC machining are essential for the production of haircutting robots. CNC machining can produce high-precision structural parts such as robotic arms and drive shafts. These parts not only need to be highly rigid but must also be manufactured with very tight tolerances to ensure the stability and accuracy of the movement system. For example, CNC machining can achieve an error margin of 0.01 mm or smaller, which is crucial for high-precision operations like hair cutting [22].

For some complex geometries, additive manufacturing (3D printing) technology can also play a role in the production of haircutting robots. With 3D printing, complex and lightweight components can be produced that are difficult to achieve with traditional machining methods. For example, additive manufacturing can be used to create lightweight plastic shells or mechanical components with built-in wiring channels, simplifying the subsequent assembly process [23].

7.3 Cost-Effective Production Strategies

While ensuring high precision and performance, effectively controlling manufacturing costs is also key to the commercialization of haircutting robots. Therefore, production processes need to combine modular design and mass production strategies to reduce unit production costs.

Modular design allows haircutting robots to use standardized components, which not only simplifies the manufacturing process but also reduces the cost of each component through mass production. Modular design also enables independent functionality in different parts of the robot, which simplifies maintenance and part replacement when malfunctions occur, further reducing long-term operational costs [24].

In addition, the production process needs to balance cost with performance. For example, in some non-critical components, low-cost plastic or composite materials can be used instead of expensive metal materials. For parts that need to bear heavy loads or have high precision requirements, higher-strength materials such as stainless steel or aluminum alloys should still be used. This strategy allows the haircutting robot to maintain overall performance while effectively controlling material and manufacturing costs.

By selecting the right materials, employing high-precision manufacturing techniques, and using modular design, haircutting robots can maximize cost-efficiency in the market. Moreover, as production technologies advance, the future introduction of automated production and intelligent manufacturing systems is expected to further reduce the manufacturing costs of haircutting robots, enabling them to reach a larger market.

8 Conclusion

In this chapter, we have thoroughly discussed several key mechanical elements for the successful design of haircutting robots. First, the mechanical structure of haircutting robots needs to achieve precise position and posture control, which can be ensured through multi-degree-of-freedom mechanical structures that provide flexibility and accuracy. We analyzed the advantages and disadvantages of Cartesian coordinate systems, rotational coordinate systems, and parallel mechanisms, and discussed the selection of degrees of freedom in design based on different application scenarios.

Next, in terms of degree of freedom design, reasonable allocation of degrees for position and posture control can effectively improve the robot's flexibility in limited spaces. Introducing redundant degrees of freedom is also an effective strategy to enhance operational safety and flexibility, helping the robot avoid the complex contours of the head and potential collision risks.

Additionally, the choice between compact and expandable design depends on the complexity of the robot's tasks. By drawing on the design principles of five-axis CNC machines, haircutting robots can use a combination of three linear axes and two rotational axes to achieve comprehensive control over position and posture. Compact design helps reduce system complexity and manufacturing costs, while expandable design provides a greater working range and flexibility, particularly for handling complex hairstyles.

Looking ahead, the trends in mechanical design improvements for haircutting robots are likely to focus on several areas. First, the combination of modular design and intelligent control algorithms will offer more possibilities for optimizing mechanical structures, enabling robots to perform complex tasks with fewer degrees of freedom. Second, as artificial intelligence and sensing technologies advance, haircutting robots will be able to better perceive their surroundings and make more precise adjustments, improving the quality and efficiency of haircuts.

The application of mechanical automation in the haircutting industry is still in its early stages. Future design improvements will further reduce system manufacturing and maintenance costs while enhancing the user experience. Through the continuous optimization of mechanical structures and control systems, haircutting robots will be able to demonstrate broader application prospects in the future market, becoming important tools in personalized services.

References

1. Merlet, J. P. (2006). *Parallel robots* (Vol. 128). Springer Science & Business Media.
2. Pedrocchi, N., Vicentini, F., Matteo, M., & Tosatti, L. M. (2013). Safe human-robot cooperation in an industrial environment. *International Journal of Advanced Robotic Systems, 10*(1), 27.
3. Kuntze, H. B. (1989). Position control of industrial robots—impacts, concepts and results. In *Robot Control 1988 (SYROCO'88)* (pp. 81–89). Pergamon.
4. Harandi, M. R. J., Khalilpour, S. A., Taghirad, H. D., & Romero, J. G. (2021). Adaptive control of parallel robots with uncertain kinematics and dynamics. *Mechanical Systems and Signal Processing, 157*, 107693.
5. Sciavicco, L., & Siciliano, B. (2012). *Modelling and control of robot manipulators*. Springer Science & Business Media.
6. Merlet, J. P., (2006) *Parallel Robots*, Kluwer Academic Publishers, 2nd Edition
7. Breńkacz, Ł, Kędra, R., Janicki, W., Maurin, A., Bagiński, P., Andrearczyk, A., & Zima, B. (2022). Research on linear actuators for active foil bearings. *Materials, 15*(16), 5694.
8. Khan, A. H., Cao, X., Li, S., Khan, A. H., Cao, X., & Li, S. (2021). Obstacle avoidance based decision making and management of articulated agents. *Management and Intelligent Decision-Making in Complex Systems: An Optimization-Driven Approach*, 1–29.
9. Kumar, S., Wöhrle, H., de Gea Fernández, J., Müller, A., & Kirchner, F. (2020). A survey on modularity and distributivity in series-parallel hybrid robots. *Mechatronics, 68*, 102367.
10. Craig, J. J., (2018). *Introduction to robotics: Mechanics and control*, Pearson Education, 4th Edition.
11. Li, S., Zhang, Y., & Jin, L. (2016). Kinematic control of redundant manipulators using neural networks. *IEEE Transactions on Neural Networks and Learning Systems, 28*(10), 2243–2254.
12. Kabir, A. M., Shah, B. C., & Gupta, S. K. (2018, August). Trajectory planning for manipulators operating in confined workspaces. In *2018 IEEE 14th International Conference on Automation Science and Engineering (CASE)* (pp. 84–91). IEEE.
13. Warkentin, A., Hoskins, P., Ismail, F., & Bedi, S. (2019). Computer aided 5-axis machining. *Computer-Aided Design, Engineering, and Manufacturing*, pp. 112–145.
14. Lin, C. Y., Jo, P. C., & Tseng, C. K. (2011). New compliance-mechanism design for small companion robots. *Journal of Intelligent & Robotic Systems, 64*, 585–601.
15. Parenti-Castelli, V., et al. (2010). Expandable robots: Design and control in complex environments. *International Journal of Robotics Research, 29*(6), 785–795.
16. Jin, L., Li, S., La, H. M., & Luo, X. (2017). Manipulability optimization of redundant manipulators using dynamic neural networks. *IEEE Transactions on Industrial Electronics, 64*(6), 4710–4720.
17. Holmes, L. A. (2022). *Hair clippers with flexing electrically adjustable blades*. U.S. Patent Application 17/546,841, filed July 14.
18. Horowitz, D., Petrelli, M. C., & Bosma, I. (2022). *U.S. Patent No. 11,267,145*. Washington, DC: U.S. Patent and Trademark Office.
19. Davis, J. R., (1998) *Aluminum and aluminum alloys*, ASM International Handbook.
20. Chawla, K. K. (2012). *Composite materials: science and engineering*. Springer Science & Business Media.
21. Gardner, L. (2005). The use of stainless steel in structures. *Progress in Structural Engineering and Materials, 7*(2), 45–55.

22. Smith, G. T. (2013). *CNC Machining Technology: Volume 3: Part Programming Techniques*. Springer Science & Business Media.
23. Rajaguru, K., Karthikeyan, T., & Vijayan, V. (2020). Additive manufacturing–State of art. *Materials today: Proceedings, 21*, 628–633.
24. Salerno, M. S., & Dias, A. V. C. (1999). Product design modularity, modular production, modular organization: The evolution of modular concepts. *Automotive Industries, 3*, 61–73.

Perception Systems and Environmental Understanding 10

With the development of robotic haircutting systems, perception systems have become one of the critical components for achieving precise haircuts. A haircutting robot's perception system is not only responsible for sensing the environment but also for accurately 3D modeling the head and hairstyle of the user, analyzing the texture and softness of the hair, and ensuring safety throughout the haircutting process. In this chapter, we will delve into 3D modeling techniques for the head and hairstyle, the application of computer vision in haircutting, and human safety recognition and obstacle avoidance technology (Fig. 1).

1 3D Modeling Techniques for the Head and Hairstyle

To achieve precise modeling of the head and hairstyle, haircutting robots rely on the collaboration of various sensors. Common sensing technologies include stereo vision, depth cameras, LiDAR, structured light, and monocular cameras. Each technology has its unique advantages and application scenarios, so multi-sensor fusion is often employed to acquire more comprehensive 3D information to improve the system's robustness.

Stereo vision is a commonly used 3D reconstruction technology. It uses two cameras from different angles to capture images of the target object, then calculates the disparity between the images to generate a depth map [1]. In haircutting robots, stereo vision can effectively capture the head's contours and the rough shape of the hairstyle. However, due to the thin and complex structure of hair, stereo vision might face challenges when dealing with individual hair strands, especially when handling curly or voluminous hairstyles, where depth information may be lost or inaccurate [2].

© The Author(s), under exclusive license to Springer Nature Switzerland AG 2025
S. Li, *Haircutting Robots*, https://doi.org/10.1007/978-3-031-84026-5_10

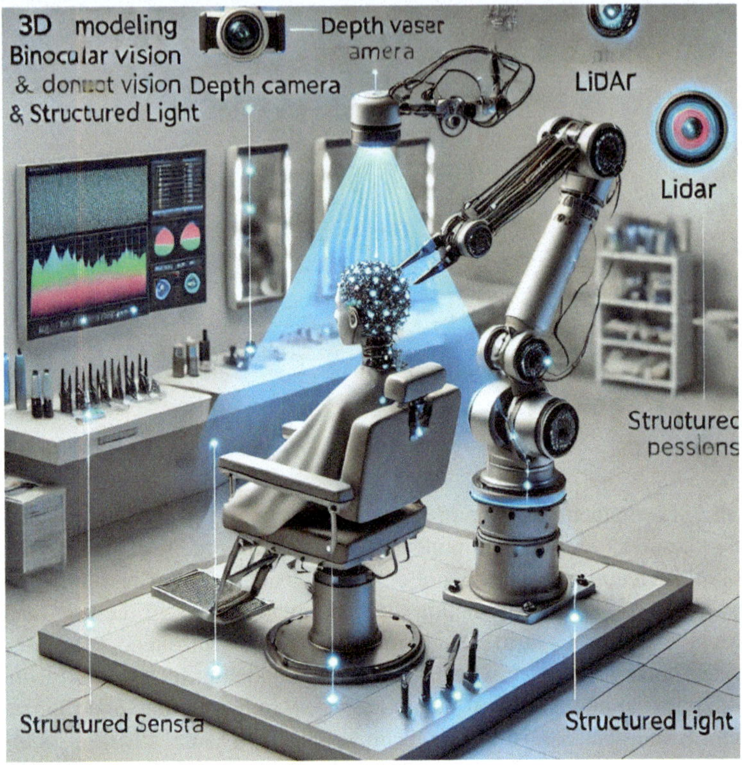

Fig. 1 Schematic of the sensing system and environmental perception of a haircutting robot

Depth cameras acquire depth information by emitting infrared light and measuring the time it takes for the light to reflect back (Time of Flight, ToF), a method widely used in devices like Microsoft's Kinect sensor [3]. For haircutting robots, depth cameras can accurately capture the overall shape of the head, making them suitable for short hair or hair closely attached to the scalp. However, due to hair's optical properties—particularly its variable infrared light reflectivity—depth cameras may struggle to capture detailed depth data for more complex hairstyles [4].

LiDAR, a technology that emits laser beams and measures the reflection time to obtain high-precision 3D data, is commonly used in autonomous driving and terrain mapping [5]. For haircutting robots, LiDAR can capture high-precision shapes of the head and hairstyle structure, especially for short or fixed hairstyles. However, LiDAR devices are expensive, and processing hair strand details may require additional post-processing steps, such as noise filtering and point cloud enhancement, to ensure accuracy in hairstyle modeling [6].

Structured light projects specific light patterns (e.g., stripes or dot matrices) onto a target surface and deduces the shape and depth of the object based on the deformation of the light. This technology has been applied in facial recognition in iPhones and boasts high precision [7]. In haircutting robots, structured light can accurately capture the details of the head and hairstyle, though the technology is sensitive to ambient lighting conditions and may require compensation in strong or dim lighting environments [8].

Monocular cameras combined with deep learning: Although monocular cameras can only capture 2D images, advances in deep learning have enabled the inference of depth information from 2D images using algorithms like Convolutional Neural Networks (CNNs) [9]. This technology works well for simpler hairstyles, especially in low-cost, low-computing power scenarios. However, for complex hairstyles, the depth inference method used by monocular cameras has limitations, often requiring integration with other sensors to enhance accuracy [10].

Compared to other static objects, 3D modeling of hair presents significant technical challenges. Hair has highly dynamic properties and a complex fiber structure, particularly for curly, voluminous, or long hairstyles that change due to factors like gravity, wind, humidity, or contact with cutting tools. Because hair is a relatively soft material, its light reflection and absorption characteristics differ greatly from those of harder surfaces like skin and bones [11]. As a result, a single sensor technology may not be able to capture all the features of hair, especially the changes and dynamic deformations of fine strands. Thus, multi-sensor fusion and algorithm optimization are key challenges in the design of haircutting robots [12].

Another challenge lies in the varying optical properties of different hair types and hairstyles. For example, oily hair reflects more light due to its smooth surface, while dry, curly hair tends to absorb more light [13]. These differences in optical properties affect the accuracy of sensors. Future research will need to improve sensor resolution and adopt machine learning and AI technologies to automatically adjust sensors to accommodate different hair types, thus obtaining more accurate 3D models [14].

In the automatic haircut process, the softness of the hair is a crucial factor affecting the robot's operational precision. By measuring hair softness, the robot can adjust the strength of the hair-cutting tool based on the hair type, ensuring precise cutting without applying excessive pressure on the user's scalp. Currently, there are several methods for measuring hair softness:

Optical Sensors: Optical sensors can analyze the reflection and absorption characteristics of hair under different lighting conditions to infer hair texture and softness. For example, soft hair typically exhibits higher light scattering, while harder hair shows more concentrated light reflection [15]. By combining optical sensors, the robot can better judge hair texture and adjust cutting strength accordingly.

Tactile Sensors: In haircutting robots, tactile sensors directly sense the force of contact between the cutting tool and the hair. Using force or pressure sensors, the system can measure the resistance changes during cutting in real-time, thus estimating hair softness. This technology is widely used in medical robots, such as the Da Vinci surgical system, which senses the contact force between surgical instruments and tissues, providing force feedback to avoid unnecessary damage to the patient [16]. Haircutting robots can borrow similar technologies to dynamically adjust cutting strength based on the feedback provided by tactile sensors, ensuring safety and comfort during haircuts.

Image Recognition and Physical Modeling: By combining image recognition with physical modeling, haircutting robots can analyze hair shape changes before and after cutting to estimate hair softness. For example, curly and straight hair exhibit different shape changes under external forces during cutting. By collecting extensive haircut data, haircutting robots can establish mapping relationships between hair texture and shape changes, predicting hair softness during the haircut process [17]. Future technological developments may integrate AI to further improve the accuracy of hair softness recognition by learning from extensive haircut data.

2 Application of Computer Vision in Haircutting

As computer vision technology rapidly advances, the visual perception module in robotic haircutting systems has become a core part of their intelligence and automation. Through computer vision, haircutting robots can precisely recognize the details of the head and hairstyle and offer personalized design and hairstyle recommendations based on user needs. Additionally, computer vision provides real-time feedback during the haircut process to ensure cutting accuracy. Below, we will discuss three major applications of computer vision in haircutting: precise hairstyle information acquisition, intelligent hairstyle recommendation, and haircut effect visualization.

Haircutting robots first need to accurately capture the details of the head and hairstyle to ensure precision during cutting. This task relies on the robust analysis capabilities of computer vision, especially through image processing and pattern recognition techniques, to extract the fine texture and shape of the hair.

The shape and texture of hair are complex and variable, especially for curly hair, layered hairstyles, and voluminous styles. To accurately capture these details, computer vision systems must capture head images from multiple angles and use advanced image processing algorithms, such as edge detection, texture analysis, and surface fitting [7]. These techniques can identify even the smallest details of the hair, ensuring that the robot can cut based on the precise shape of the hairstyle. Additionally, the use of high-resolution cameras and 3D depth sensing can further enhance the visual system's ability to resolve complex hairstyle structures [8].

In the haircutting process, the computer vision system is not merely an observation tool; it also captures the relative position of the cutting tool and the hair in real time and adjusts the cutting path through a rapid feedback mechanism to ensure precision. Computer vision can continuously collect images, analyzing the dynamic changes of the hair during the cutting process, such as the movement or deformation of the hair [9]. This real-time feedback significantly improves the robot's operational accuracy, preventing issues like asymmetrical hairstyles or excessive cutting.

In today's era, where personalized services are integral to user experience, computer vision can also recommend suitable hairstyles for users through facial recognition and big data analysis. Each person's face has unique characteristics, and an appropriate hairstyle can enhance their appearance. Based on this need, the computer vision system analyzes the user's facial features and, combined with a hairstyle database and AI algorithms, provides intelligent hairstyle recommendations (Fig. 2).

Using facial feature point detection technology, computer vision can precisely extract the user's facial geometry, including face shape, cheekbone position, and jawline angle [10]. These identified facial features help the system determine the user's face type, such as round, square, or oval. Based on this, the system can select a hairstyle that matches the user's face shape from an established hairstyle database [11].

By combining computer vision technology with big data analysis, the system can recommend personalized hairstyles based on various factors, including the user's age, gender, face shape, and even fashion trends. AI algorithms, such as

Fig. 2 Facial feature point detection technology extracts the user's facial geometry to match a suitable hairstyle

Convolutional Neural Networks (CNN) and Recurrent Neural Networks (RNN), excel at processing large datasets, enabling them to uncover hidden matching rules from a vast amount of user data and hairstyle designs, providing new users with hairstyles that best suit their individual characteristics [12]. This recommendation not only relies on face shape analysis but also integrates hairstyle trends and personal preferences, further enhancing the recommendation's accuracy and user satisfaction [13].

When providing intelligent services, users often want to preview the results before the actual haircut. To meet this demand, augmented reality (AR) and 3D modeling technologies have been widely applied. With these technologies, haircutting robots can showcase various hairstyle outcomes before cutting, allowing users to make informed choices based on the previews, improving interactivity and overall service experience.

AR technology allows virtual hairstyles to be superimposed onto the user's real image in real time, creating a preview effect [14]. By using the camera to capture the user's face, AR technology can overlay different 3D hairstyle models onto the user's head in real time, simulating the haircut results. Users can interact with the system's interface to select different hairstyles and instantly view how each one looks on their head. This visual preview feature greatly enhances users' ability to predict the results of the haircut, reducing dissatisfaction after the haircut [15].

Computer vision, combined with 3D modeling technology, can provide users with a multi-angle preview of different hairstyles. By constructing a 3D model of the user's head, the system can simulate how various hairstyles will appear from different angles, such as front, side, and back views [16]. This not only enhances users' decision-making abilities but also enables the hairdresser or robot to better understand user needs and make adjustments based on the previewed effect. In the future, with the further development of AR and 3D technology, haircutting robots could even offer real-time before-and-after comparisons of haircuts, providing users with more personalized and interactive haircutting experiences [17].

3 Sensors for Ensuring Human Safety and Obstacle Avoidance

In haircutting robot applications, user safety is a primary concern. Haircutting robots must not only precisely complete hairstyle trimming tasks but also ensure that they do not cause any harm to the user. To achieve this, sensors play a crucial role in the processes of safety recognition and obstacle avoidance. By using various types of sensors, haircutting robots can automatically detect obstacles in the complex environment, plan safe paths, and recognize human safety while controlling contact pressure. Below, we'll explain in detail the implementation of obstacle avoidance path planning, safe contact force control, and the integration of surgical robot technology.

During a haircut, the robot needs to skillfully avoid obstacles on the head (such as ears or different hair layers) while ensuring that its mechanical arm or cutting tool does not collide with other parts of the user's body. To accomplish this, obstacle avoidance path planning becomes a critical part of haircutting robot design. The combination of computer vision and sensor systems enables the robot to perceive its surroundings and generate a real-time updated 3D model for path planning (Fig. 3).

Using stereo vision technology, haircutting robots can construct a real-time 3D model of the user's head [18]. This model includes not only the overall contour of the head but also captures the specific shape of the hairstyle. By sensing these details, the robot can adjust the movement of the hair-cutting tool to ensure it avoids any areas that should not be touched.

Fig. 3 Using technologies like range measurement to detect the safe distance between the robot and the customer in real time for safety assurance

During obstacle avoidance, the robot must detect all objects in its environment, which includes not only the user's head but also any physical objects that may appear during the haircut (such as chairs or tools). Obstacles can be detected in real-time using sensors such as LiDAR, depth cameras, or infrared sensors [19]. Once an obstacle is identified, path planning algorithms, such as the A* algorithm or Rapidly-exploring Random Trees (RRT), calculate a safe path to avoid the obstacle [20]. The A* algorithm searches the grid for the shortest safe path from the robot's current position to the target, while RRT quickly generates an avoidance path in complex environments, making it suitable for real-time planning in dynamic settings [21]. These path-planning algorithms ensure that the haircutting robot does not collide with any obstacles throughout the operation, maximizing user safety.

During the haircutting process, the contact force between the hair-cutting tool and the scalp is crucial for ensuring safety. To prevent the cutting tool from exerting excessive pressure on the scalp, the robot needs to precisely monitor and control the contact force. By using force sensors and motor control systems, the robot can measure the force applied by the cutting tool in real-time and automatically adjust the force to ensure a safe and comfortable haircut.

Force sensors are a core component of contact control for robots. They can detect the force applied by the cutting tool when it touches the head and provide feedback to the control system [22]. For example, if the sensor detects that the force applied by the tool exceeds a preset safety threshold, the system automatically reduces the tool's speed or pressure to prevent excessive force on the user's scalp. In addition to force sensors, the haircutting robot can estimate the applied force through motor current monitoring systems [23]. By measuring the current consumed by the motor, the system can infer changes in resistance when the cutting tool contacts the hair, allowing further force adjustment.

In addition to directly measuring contact force, infrared sensors and camera-based sensors can detect micro-deformations in the user's skin during the haircut [24]. For example, infrared sensors can monitor the distance between the cutting tool and the scalp and estimate whether the tool is applying excessive pressure by analyzing small deformations on the skin surface [25]. This non-contact monitoring method complements the force sensors, further enhancing safety.

Surgical robots have been widely applied in the medical field, achieving significant success in force sensing and safety feedback technologies during minimally invasive surgeries. Haircutting robots can borrow these technologies, especially in terms of contact force control and precise operations, to further improve safety and accuracy.

The Da Vinci surgical system is currently one of the most advanced surgical robots, and one of its key technologies is the ability to sense the contact force between surgical instruments and tissues in real-time [26]. This force feedback system, using high-precision force sensors and tactile feedback, ensures that every step of the surgical procedure is carried out within a safe range. Haircutting robots can adopt similar technology by integrating force sensing into the design of the cutting tool, ensuring that no harm is caused to the user's skin or scalp during the haircut [27].

Minimally invasive surgery requires surgical robots to perform highly precise operations in confined spaces while avoiding damage to surrounding tissues [28]. Haircutting robots can learn from this, especially when performing precise trimming near sensitive areas like the ears. The coordination between force sensors and path-planning algorithms helps the robot execute accurate cuts without causing harm to the user [29].

By incorporating these mature technologies from surgical robots, haircutting robots can not only improve operational precision but also significantly enhance safety during complex haircutting tasks (Fig. 4).

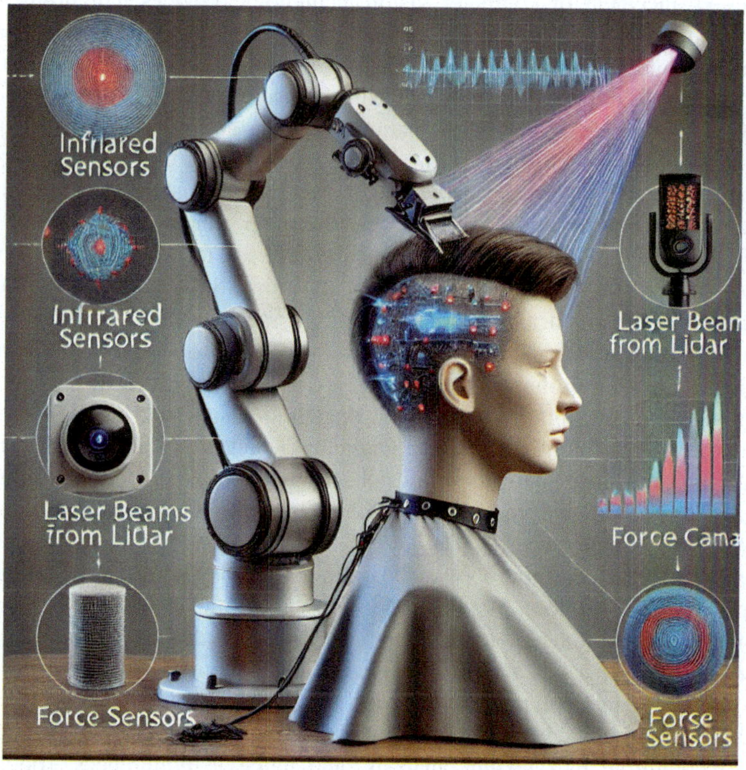

Fig. 4 The haircutting robot achieves complex environmental modeling through the collaborative work of multiple sensors, such as infrared sensors, LiDAR, depth cameras, and force sensors

4 Conclusion and Future Outlook

The integration of several key technologies has laid the foundation for haircutting robots to achieve intelligent, precise, and safe haircutting services. From sensor technology to computer vision, to path planning and contact force control, the entire system's collaborative operation enables haircutting robots to efficiently perform real-time sensing and adjustments.

The integration of sensors is at the core of the haircutting robot system's intelligence. From 3D models constructed using technologies like stereo vision and depth cameras to force sensors and infrared sensors monitoring safety, each sensing module forms a unified perception system through data fusion. This collaboration of multiple sensors not only enhances the robot's ability to perceive the head and hairstyle but also significantly improves the safety and precision of the haircutting process [18, 19].

The advancement of computer vision technology provides haircutting robots with intelligent perception capabilities, enabling them to capture hairstyle details in real-time and make dynamic adjustments. Moreover, the integration of real-time visual feedback and path-planning algorithms ensures precise robot operations in complex environments. This combination of vision and control systems greatly enhances the efficiency of haircutting robots and improves user experience [30].

The integration of force sensors and motor control systems ensures that the contact force between the cutting tool and the scalp remains within a safe range during the haircut. Data from the sensors are fed back to the robot via the motor control system, allowing real-time adjustment of force during the haircut to ensure the user is not harmed. This force feedback system, inspired by force-sensing technology used in surgical robots, has brought significant advances in operational safety for haircutting robots [31, 32].

Through the collaboration of sensing, modeling, vision, and safety recognition technologies, haircutting robots have gradually achieved a high degree of intelligence. This integrated innovation has not only improved the robot's performance but has also further advanced human–machine interaction technologies, enhancing the overall user experience.

In the future, haircutting robots will continue to evolve towards more intelligent and personalized services. With advancements in artificial intelligence, machine learning, and sensing technologies, haircutting robots will make breakthroughs in understanding user needs and providing adaptive haircutting services.

With the progress of big data and AI technologies, future haircutting robots will be better equipped to understand users' personalized needs. By integrating facial recognition, hair texture analysis, and hairstyle preferences, robots will offer customized haircutting services for each user. As robots continually learn and optimize, they will accumulate more data, enabling them to offer even more personalized hairstyle recommendations and designs [11, 33].

As perception systems and environmental understanding technologies further develop, haircutting robots will work efficiently in more complex environments. In the future, these robots may not be limited to haircutting tasks but could expand into more intricate beauty services such as hairstyling and coloring. With enhanced perception and understanding abilities, haircutting robots will autonomously navigate users' environments and automatically adjust and optimize services based on user needs [27].

Safety has always been one of the focal points in the design of haircutting robots. With the continued optimization of force-sensing technology and obstacle avoidance algorithms, future haircutting robots will be able to avoid collisions with the user's head more flexibly and accurately. Additionally, with improved human–machine interaction interfaces, users will be able to interact with the robot more easily and intuitively, enhancing the overall user experience [17, 34] (Fig. 5).

As virtual reality (VR) and augmented reality (AR) technologies continue to develop, haircutting robots will leverage these technologies to provide users with virtual hairstyle previews. Through real-time 3D modeling, users will be able to

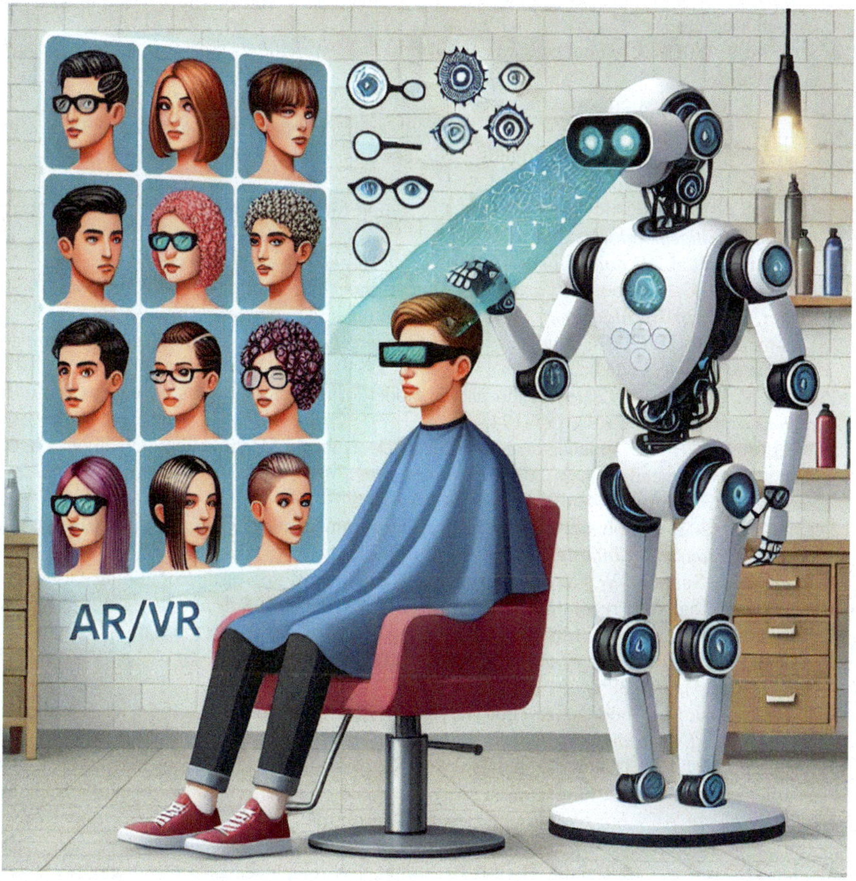

Fig. 5 Customers can preview their appearance with a selected hairstyle through VR-generated virtual previews

preview the effects of different hairstyles before a haircut, helping them make more informed decisions [15, 35]. This interactive mode not only enhances user engagement but also introduces a new experience to haircutting services.

By integrating and collaborating across multiple technologies, haircutting robots have made significant advancements. In the future, as sensing technology, computer vision, AI, and force-sensing technology continue to evolve, haircutting robots will become more intelligent, efficient, and safe. By deeply understanding user needs and improving operational precision, haircutting robots are expected to become essential tools in the beauty industry, promoting the widespread adoption of personalized and automated haircutting services.

References

1. Hartley, R., & Zisserman, A. (2003). *Multiple view geometry in computer vision.* Cambridge University Press.
2. Szeliski, R. (2010). *Computer vision: Algorithms and applications.* Springer Science & Business Media.
3. Zhang, Z. (2012). Microsoft Kinect sensor and its effect. *IEEE Multimedia, 19*(2), 4–10.
4. Liu, F., Shen, C., Lin, G., & Reid, I. (2015). Learning depth from single monocular images using deep convolutional neural fields. *IEEE Transactions on Pattern Analysis and Machine Intelligence, 38*(10), 2024–2039.
5. Chen, L. (2020). LIDAR-based 3D object detection for autonomous driving. *IEEE Transactions on Intelligent Vehicles, 5*(4), 682–692.
6. Nagami, T., et al. *3D point cloud denoising using a deep learning approach.*
7. Gonzalez, R. C., & Woods, R. E. (2017). *Digital image processing* (4th ed.). Pearson.
8. Parker, J. R. (2010). *Algorithms for image processing and computer vision.* Wiley.
9. Dickmanns, E. D. (2007). *Dynamic vision for perception and control of motion.* Springer Science & Business Media.
10. Blanz, V., & Vetter, T. (2019). A morphable model for the synthesis of 3D faces. *ACM Transactions on Graphics, 29*(4), 187–194.
11. Weerasinghe, H., & Vidanagama, D. (2020, December). Machine learning approach for hairstyle recommendation. In *2020 5th international conference on information technology research (ICITR)* (pp. 1–4). IEEE.
12. Kim, T. K. (2023). Implementation of hair style recommendation system based on big data and deepfakes. *Journal of Internet of Things and Convergence, 9*(3), 13–19.
13. Pasupa, K., Sunhem, W., & Loo, C. K. (2019). A hybrid approach to building face shape classifier for hairstyle recommender system. *Expert Systems with Applications, 120*, 14–32.
14. Billinghurst, M., Kato, H. (2019). *Collaborative mixed reality.* MIT Press.
15. Azuma, R. (2019). A survey of augmented reality. *Presence: Teleoperators and Virtual Environments, 6*(4), 355–385.
16. Lyu, Q., Chai, M., Chen, X., & Zhou, K. (2020). Real-time hair simulation with neural interpolation. *IEEE Transactions on Visualization and Computer Graphics, 28*(4), 1894–1905.
17. Singh, S., Mali, H., Thakur, V., & Lovalekar, S. (2024, July). HairVise: A machine learning, augmented reality and artificial intelligence integrated personalized hair care recommendation framework. In *2024 8th international conference on inventive systems and control (ICISC)* (pp. 288–294). IEEE.
18. Ham, H., Wesley, J., & Hendra, H. (2019). Computer vision based 3D reconstruction: A review. *International Journal of Electrical and Computer Engineering, 9*(4), 2394.

19. Ghorpade, D., Thakare, A. D., & phode, S. (2017, August). Obstacle detection and avoidance algorithm for autonomous mobile robot using 2D LiDAR. In *2017 international conference on computing, communication, control and automation (ICCUBEA)* (pp. 1–6). IEEE.
20. Hart, P. E., Nilsson, N. J., & Raphael, B. (1968). A formal basis for the heuristic determination of minimum cost paths. *IEEE Transactions on Systems Science and Cybernetics, 4*(2), 100–107.
21. LaValle, S. M. (2019). *Rapidly-exploring random trees: A new tool for path planning, TR 98–11*. Computer Science Department, Iowa State University.
22. Cao, M. Y., Laws, S., & y Baena, F. R. (2021). Six-axis force/torque sensors for robotics applications: A review. *IEEE Sensors Journal, 21*(24), 27238–27251.
23. Wang, Z., Guo, J., Guo, S., & Song, Y. (2022, August). Study on force feedback for virtual vascular interventional surgical system based on multi-data fusion. In *2022 IEEE international conference on mechatronics and automation (ICMA)* (pp. 1525–1530). IEEE.
24. Zang, H., Zhang, X., Zhu, B., & Fatikow, S. (2019). Recent advances in non-contact force sensors used for micro/nano manipulation. *Sensors and Actuators A: Physical, 296*, 155–177.
25. Joasoon, E. (2010). *The tactile feedback device for multi-touch user interfaces*. TUT Press.
26. Thai, M. T., Phan, P. T., Hoang, T. T., Wong, S., Lovell, N. H., & Do, T. N. (2020). Advanced intelligent systems for surgical robotics. *Advanced Intelligent Systems, 2*(8), 1900138.
27. Okamura, A. M. (2020). Haptic feedback in robot-assisted minimally invasive surgery. *Current Opinion in Urology, 20*(2), 109–114.
28. Wang, Y., Uecker, D., Laby, K. P., Wilson, J. D., Jordan, C. S., Wright, J. W., & Ghodoussi, M. (2008). U.S. Patent Application No. 12/104, 271.
29. Fei, B., Ng, W. S., Chauhan, S., & Kwoh, C. K. (2001). The safety issues of medical robotics. *Reliability Engineering & System Safety, 73*(2), 183–192.
30. Hart, P. E., Nilsson, N. J., & Raphael, B. (2020). A formal basis for the heuristic determination of minimum cost paths. *IEEE Transactions on Systems Science and Cybernetics, 4*(2), 100–107.
31. Tegin, J., & Wikander, J. (2005). Tactile sensing in intelligent robotic manipulation–a review. *Industrial Robot: An International Journal, 32*(1), 64–70.
32. Ortmaier, T., Deml, B., Kübler, B., Passig, G., Reintsema, D., & Seibold, U. (2007). Robot assisted force feedback surgery. *Advances in Telerobotics*, 361–379.
33. Deldjoo, Y., Nazary, F., Ramisa, A., Mcauley, J., Pellegrini, G., Bellogin, A., & Noia, T. D. (2023). A review of modern fashion recommender systems. *ACM Computing Surveys, 56*(4), 1–37.
34. Camarillo, D. B., Krummel, T. M., & Salisbury, J. K., Jr. (2004). Robotic technology in surgery: Past, present, and future. *The American Journal of Surgery, 188*(4), 2–15.
35. Ward, K., Bertails, F., Kim, T. Y., Marschner, S. R., Cani, M. P., & Lin, M. C. (2007). A survey on hair modeling: Styling, simulation, and rendering. *IEEE Transactions on Visualization and Computer Graphics, 13*(2), 213–234.

Control Algorithms and Motion Planning 11

1 Algorithm Modeling of Haircutting Actions

The core control of a haircutting robot lies in the motion planning and execution of its manipulator arm. As a mechanical system with high degrees of freedom, each hair-cutting action requires precise control algorithms to ensure smoothness and accuracy of movement. This section focuses on the algorithmic modeling of haircutting actions, covering inverse kinematics, neural network–based control strategies, and other intelligent control methods to ensure the robot's flexibility and safety during the haircutting process.

1.1 Inverse Kinematics Modeling of the Manipulator Arm

Inverse Kinematics (IK) is crucial in manipulator arm control. Its goal is to determine the angles or displacements of each joint by specifying the desired position and orientation of the end effector (e.g., haircutting tools), ensuring that the end effector precisely reaches the specified location. In haircutting robots, IK is used to determine the exact position and angle of the haircutting tools in space to achieve precise hair trimming (Fig. 1).

Each joint of the manipulator has multiple degrees of freedom, requiring complex control over the end effector's position and orientation in a multi-degree-of-freedom space. Inverse kinematics maps the end effector's spatial position to the joint space and solves it through mathematical models.

For simpler manipulators, such as non-redundant systems, IK can be directly solved analytically. Analytical methods are usually based on geometric relationships, calculating the precise angle of each joint through the manipulator's lengths, angles, and spatial geometric properties. These methods are suitable for manipulators with simple structures, offering fast and efficient computations.

© The Author(s), under exclusive license to Springer Nature Switzerland AG 2025

S. Li, *Haircutting Robots*, https://doi.org/10.1007/978-3-031-84026-5_11

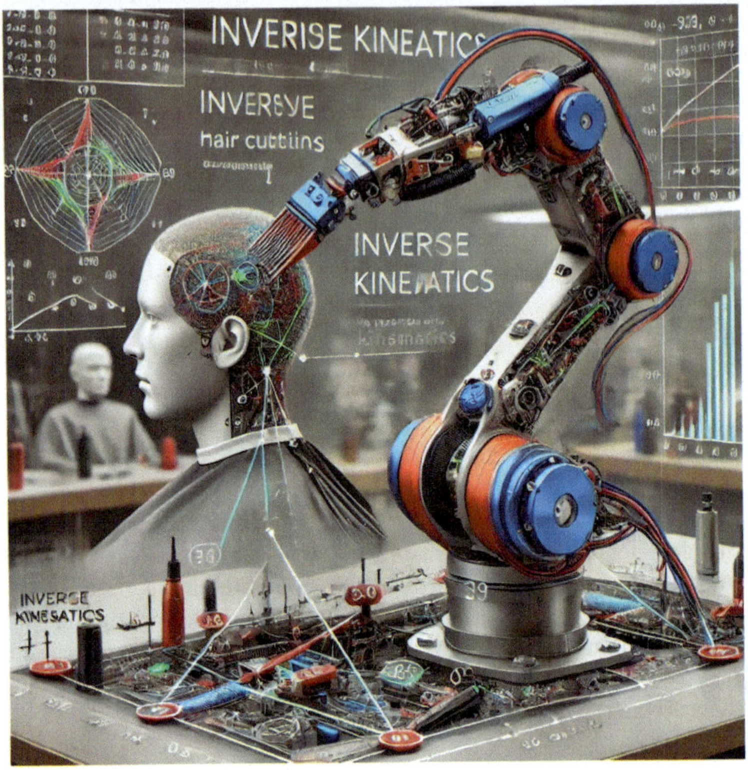

Fig. 1 The core problem inverse kinematics needs to solve: calculating the corresponding joint angles of the manipulator after planning the end effector's position

However, as the manipulator's complexity increases—especially when it has redundancy (i.e., more degrees of freedom than necessary)—the IK problem no longer has a unique solution. Redundancy allows the manipulator to have multiple joint configurations to reach a target position, providing additional flexibility to optimize other performance goals such as obstacle avoidance, joint wear reduction, and energy minimization. Therefore, numerical methods are typically used to solve the IK problem for redundant manipulators. Numerical methods use iterative calculations to approximate the ideal solution, making them suitable for complex, high-degree-of-freedom mechanical systems.

1.2 Optimization Solution Methods

For redundant manipulators, optimization solution methods are widely applied in solving inverse kinematics to ensure efficient and accurate target positioning under multiple degrees of freedom. One commonly used optimization method is the least squares–based approach, which minimizes the deviation between the target and actual positions to precisely position the manipulator's end effector. The least squares method is particularly useful for solving systems with multiple variables, especially in complex scenarios requiring simultaneous adjustment of multiple joints.

Another prevalent optimization technique is the Jacobian pseudoinverse method. The Jacobian matrix describes the relationship between the end effector's velocity and the velocities of each joint, serving as a linear approximation of the kinematic equations. By computing the pseudoinverse of the Jacobian matrix, the IK problem can be solved quickly. The Jacobian pseudoinverse method offers efficient computation, making it especially suitable for real-time control scenarios [1]. In haircutting robots, real-time responsiveness is crucial; the robot must quickly adapt to user head movements and hairstyle adjustments. Therefore, the Jacobian pseudoinverse method excels in controlling the smoothness and precision of movements during the haircutting process.

Moreover, the Jacobian pseudoinverse method effectively handles redundancy issues. By optimizing redundant joint degrees of freedom, the robot can reach the target position while ensuring the manipulator's posture remains stable during haircutting—a critical factor in avoiding collisions between the tools and the user's head. In practice, this method continuously adjusts the manipulator's posture, allowing the robot to flexibly adapt to different users' head shapes and hairstyle requirements, ensuring precise and smooth haircutting actions.

1.3 Redundancy Handling

Redundancy refers to a manipulator having more degrees of freedom than the task requires. This extra flexibility allows the manipulator to adjust additional degrees of freedom to achieve other objectives while completing the primary task. In haircutting robots, redundancy handling is essential as it permits the robot to optimize joint movements to avoid collisions or fulfill other control goals while ensuring the end effector performs the haircutting task precisely (Fig. 2).

The primary advantage of redundancy handling is increased flexibility. While reaching the target position, the manipulator can adjust its posture using the extra degrees of freedom. For example, when cutting a user's hair, the haircutting tool's end position must align precisely with the hair, but the manipulator can adjust other joints to avoid sensitive areas like ears and nose, preventing accidental collisions. This flexibility significantly enhances operational safety.

Additionally, redundancy can optimize energy consumption and reduce joint load. In high-frequency and long-duration operations, energy consumption and joint wear are critical considerations. Redundancy optimization allows the robot to select a joint configuration that completes the task while minimizing energy use. This strategy extends the manipulator's lifespan and improves system efficiency. For instance, during prolonged haircutting tasks, redundancy optimization ensures the robot performs operations with minimal energy consumption, enhancing user experience and reducing maintenance costs [2].

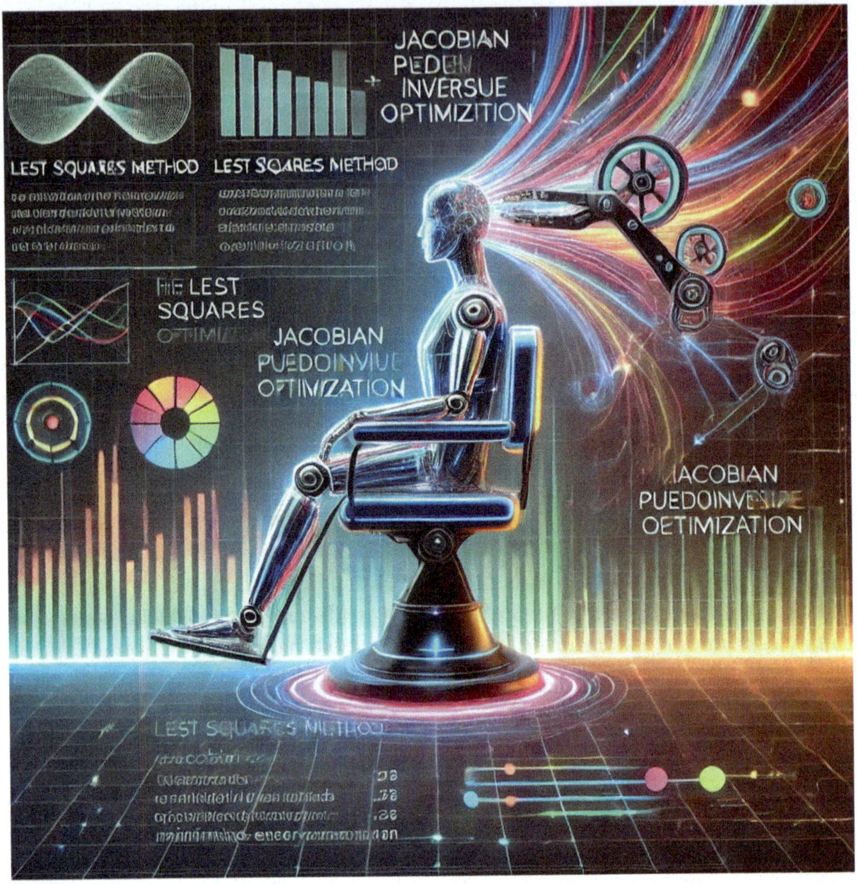

Fig. 2 Haircutting robots use optimization methods to control redundant manipulators, fully utilizing extra degrees of freedom to enhance motion flexibility and optimality

Redundancy handling also improves the robot's adaptability in space-constrained environments. For complex head structures and hairstyle requirements, extra degrees of freedom help the robot execute precise motion control within limited operating spaces, ensuring high accuracy and reliability of haircutting actions. By introducing additional optimization objectives, the haircutting robot not only efficiently completes tasks but also autonomously adjusts and adapts in complex environments, demonstrating strong flexibility and adaptability in practical applications.

1.4 Neural Network–Based Control Strategies

With advancements in machine learning and neural networks, inverse kinematics solutions based on neural networks have become a popular research direction. Compared to traditional analytical and numerical methods, neural networks can handle complex nonlinear problems, performing exceptionally well in dynamic and variable environments. The haircutting robot operates in a highly dynamic and uncertain environment, requiring high precision in haircutting actions; thus, neural network control strategies offer new possibilities.

Recurrent Neural Networks (RNNs) and Convolutional Neural Networks (CNNs) have significant advantages in processing dynamic sequence data and spatial information. RNNs excel at handling time-series data, predicting future states based on past inputs. In the haircutting process, RNNs can predict changes in haircutting actions in real-time and adjust the manipulator's motion trajectory, ensuring continuity and smoothness [3]. Conversely, CNNs are suitable for processing complex spatial structures. In haircutting actions, CNNs can better recognize hair shapes and contours, optimizing the haircutting path.

Dynamic neural networks are control strategies capable of real-time self-adjustment, particularly suitable for manipulator control. The haircutting robot can automatically learn from real-time sensor data and visual feedback, correcting errors in haircutting actions. This adaptive capability enables the robot to handle unforeseen situations—such as changes in hair thickness or user movements—ensuring operational stability and high precision [4].

1.5 Other Intelligent Control Methods

Beyond neural networks, other intelligent control methods are widely applied in controlling haircutting robots, offering diverse solutions to address uncertainties and complexities in various environments.

Fuzzy Logic Control is effective for handling uncertainties and imprecise information. Unlike traditional deterministic control, fuzzy logic allows for continuous control in highly uncertain environments, making it suitable for fine adjustments during haircutting. Through fuzzy logic, the robot can dynamically adjust the speed and posture of the haircutting tools based on the fuzziness of input data, ensuring the gentleness and precision of actions [5].

Genetic Algorithms are global optimization techniques based on natural selection and genetic variation, suitable for solving multi-objective optimization problems. In the motion planning of haircutting robots, genetic algorithms can automatically generate optimal haircutting paths by simulating evolutionary processes. They are particularly effective in addressing motion planning for redundant manipulators, achieving efficient control and obstacle avoidance while ensuring the smoothness and accuracy of the haircutting path [6].

2 Path Planning and Fine Motion Control (Planning and Feedforward)

In haircutting robots, path planning and motion control are critical to ensuring a smooth, efficient, and safe haircutting process. By employing effective path planning and motion control algorithms, the robot can achieve dynamic obstacle avoidance and trajectory optimization in complex environments. This section discusses classical path planning methods, optimization-based trajectory planning techniques, and the applications of neural networks and deep learning in path planning.

2.1 Classical Path Planning Methods

Classical path planning methods are typically used for generating macroscopic paths, ensuring that the manipulator can effectively avoid obstacles during the haircutting process and complete predetermined tasks. These methods are mainly based on graph search algorithms and random sampling algorithms, applied to the robot's path planning from the starting point to the endpoint.

The A* algorithm and Dijkstra's algorithm are two classic discrete methods in the field of path planning. They discretize the robot's motion space into grids or graphs, incrementally computing the shortest path from the starting point to the target point. The A* algorithm combines heuristic search by estimating the cost from the current node to the target node, accelerating the search process. It is suitable for path generation in static or mildly dynamic environments for haircutting robots [7]. Dijkstra's algorithm is a non-heuristic global optimal algorithm that guarantees finding the shortest path. Although it has higher computational complexity, it is still widely applied in path planning problems [8].

In haircutting robots, these path planning algorithms are mainly used for generating macroscopic paths, such as determining the movement path of the haircutting tools around the head. This ensures that the manipulator can smoothly move from one area to another, avoiding collisions with sensitive areas like the user's ears and forehead.

The Rapidly-Exploring Random Tree (RRT) is a random sampling path planning algorithm suitable for dynamic environments. Unlike the A* and Dijkstra algorithms, RRT does not require discretizing the environment but incrementally expands feasible motion paths through random sampling. This makes RRT highly efficient when handling path planning in high-dimensional spaces [9]. RRT-Connect is an improved version of the RRT algorithm that adopts a bidirectional expansion strategy, further accelerating the speed of path generation.

In haircutting robots, RRT and RRT-Connect are suitable for the micro-motion planning of the haircutting tools, especially when avoiding obstacles and dynamically adjusting paths. These algorithms can generate continuous and smooth motion trajectories, ensuring precise operations of the manipulator near complex head structures.

Besides classical path planning algorithms, optimization-based trajectory planning methods also play a significant role in the motion control of haircutting robots. These methods generate optimal motion trajectories through optimization algorithms, ensuring the smoothness and efficiency of haircutting actions.

Dynamic programming is an optimization method for solving globally optimal paths and is widely used in robot motion planning. By recursively decomposing the problem into multiple subproblems, dynamic programming can ensure an optimal solution while improving computational efficiency. In haircutting robots, dynamic programming can ensure that the haircutting tools move along a globally optimal path throughout the operation, avoiding local suboptimal traps [10].

2.2 Trajectory Smoothing Based on Bézier Curves and Spline Curves

In haircutting actions, the smoothness of the cutting path directly affects the quality of the haircut. Bézier curves and spline curves are two commonly used geometric methods for generating continuous and smooth motion trajectories. Bézier curves generate smooth curves through control points, suitable for defining the cutting path of haircutting tools. Spline curves offer greater flexibility, allowing adjustment of the curve shape based on local conditions, ensuring that the haircutting tools can operate smoothly along the trajectory, similar to the slicing algorithms used in 3D printing [11] (Fig. 3).

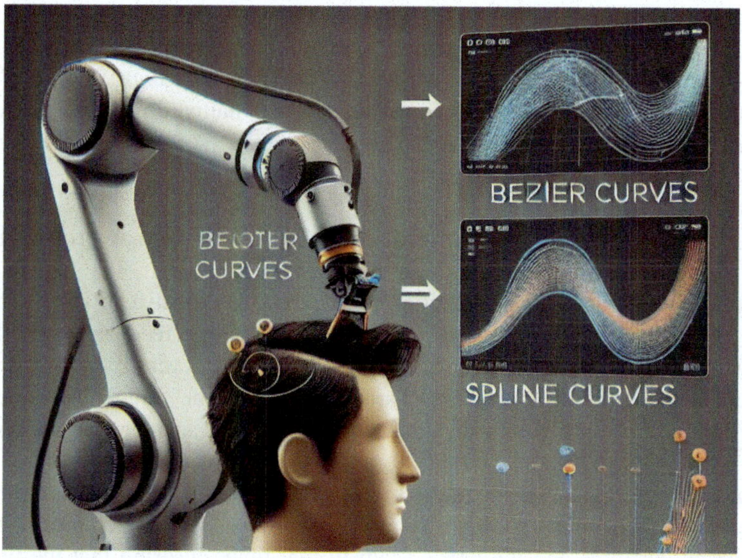

Fig. 3 Techniques like Bézier curves enable smooth joint movements of the manipulator, ensuring actions proceed smoothly

2.3 Neural Network and Deep Learning Methods

With the rise of machine learning, neural networks and deep learning methods have demonstrated powerful capabilities in the field of path planning. By learning complex nonlinear mappings, neural networks can automatically generate and optimize robot motion paths, making them particularly suitable for real-time motion planning in unstructured and dynamic environments.

Convolutional Neural Networks (CNNs) excel at processing spatial data. In haircutting robots, CNNs can generate optimal haircutting paths by analyzing three-dimensional models of the head and hairstyle. Recurrent Neural Networks (RNNs), on the other hand, are proficient at handling time-series data and can dynamically adjust the path based on real-time feedback during the haircutting process, ensuring that the motion trajectory of the haircutting tools is smooth and accurate [12].

The advantage of neural networks lies in their adaptive learning capabilities. By continuously learning historical data from the haircutting process, the haircutting robot can gradually optimize its motion strategies, achieving more intelligent and efficient haircutting operations.

3 Real-Time Feedback and Adjustment Mechanisms (Feedback and Safety Assurance)

In haircutting robots, real-time feedback and adjustment mechanisms are crucial to ensure that the robot can quickly respond and adjust according to continuously changing environments and conditions. This section will detail various aspects of the real-time feedback system, including force feedback control, visual feedback, obstacle avoidance and trajectory correction, and neural network–based adaptive control. Through these feedback mechanisms, the haircutting robot can ensure operational safety, precision, and continuity.

3.1 Force Feedback Control and Contact Force Adjustment

In the haircutting process, controlling the contact force between the haircutting tools and the user's scalp is vital. The purpose of force feedback control is to monitor the force applied by the tools in real-time to prevent excessive pressure that could cause discomfort or even harm to the user. The core of the force feedback control system lies in efficient sensor monitoring and feedback adjustment (Fig. 4).

Force sensors and torque sensors are among the most important feedback devices in robot control systems. They can monitor the contact force between the haircutting tools and the user's scalp in real-time, ensuring that the pressure applied remains within a safe range during the haircutting process. Force sensors can precisely capture minute pressure changes, while torque sensors can monitor

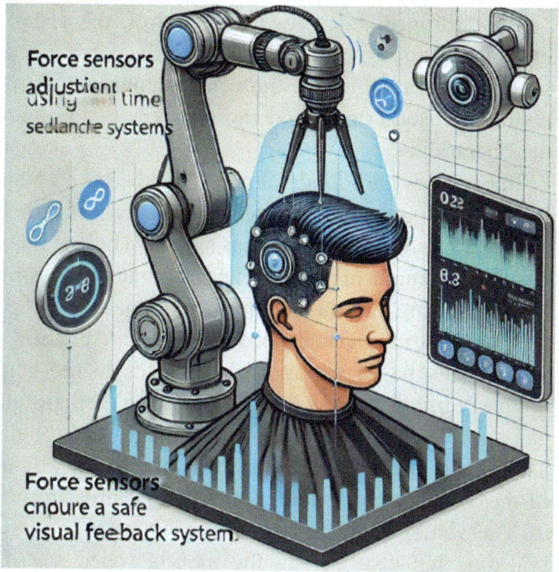

Fig. 4 Some obstacles or unexpected events cannot be predicted during manipulator operation; safety is ensured through online feedback mechanisms

the rotational torque of the tools. By real-time monitoring of the applied force, the robot can adjust its operating force accordingly to avoid causing harm to the scalp.

For example, when the force sensor detects that the force applied by the tools is too great, the robot control system will automatically reduce the speed or force of the tools to ensure that the user's scalp is not subjected to excessive pressure. This force feedback control technology has been widely used in surgical robots and other precision operation robots and has been adopted in haircutting robots [13].

3.2 PID Control Algorithms and Force Feedback Control

PID control (Proportional-Integral-Derivative control) is a classic feedback control algorithm widely used in industrial automation. In force feedback control systems, the PID control algorithm can adjust the actions of the haircutting tools in real-time, automatically adjusting the force based on feedback data from the force sensors. The PID control algorithm is characterized by fast response speed and high precision, making it suitable for precise force control during the haircutting process.

The basic principle of PID control is to perform feedback adjustments based on real-time error values (the difference between the target value and the actual value). The force sensor monitors the pressure of the tools in real-time and compares it with a preset safe range. When the actual applied force exceeds the set range, the PID control algorithm immediately makes adjustments to ensure the smoothness and safety of the haircutting operation [14].

In addition to force feedback control, the visual feedback system plays a key role in haircutting robots. Visual feedback provides comprehensive dynamic data support by obtaining real-time environmental information, three-dimensional models of the user's head, and the progress of the haircut, ensuring the precision of the operation.

3D sensors and depth cameras are the core components of the visual feedback system. Through these devices, the haircutting robot can obtain real-time three-dimensional models of the user's head. These models help the robot recognize the shape and length of the hair and accurately determine the current position of the tools, ensuring that each action is completed as planned.

The real-time visual feedback system enables the robot to perform online adjustments based on dynamic changes in the head model. For example, when the user slightly moves their head during the haircutting process, the depth camera can immediately capture this change and feed the data back to the control system. The system processes these data and automatically adjusts the trajectory and angle of the tools, ensuring the continuity and precision of the haircut [15].

The visual feedback system can not only monitor the haircutting progress in real-time but also correct the motion trajectory of the tools based on real-time data. For example, when the system detects that the cutting path deviates from the preset trajectory, the control system will immediately adjust the motion path to ensure the expected haircutting effect. This online adjustment mechanism allows the robot to continuously correct its actions, avoiding misoperations or missed areas.

Haircutting robots not only require precise cutting capabilities but must also avoid obstacles in complex environments. Obstacle avoidance technology ensures that the robot can avoid sensitive areas such as the user's ears and nose through sensor fusion and feedback control while maintaining the smoothness of the trajectory.

In haircutting robots, sensor fusion technology provides comprehensive environmental perception by integrating data from different sensors (such as LiDAR, ultrasonic sensors, and visual systems). LiDAR can scan the user's head shape with high precision, generating real-time 3D environmental models, while ultrasonic sensors detect nearby obstacles. After fusion processing, these data provide precise obstacle location information, ensuring the safety of the cutting actions [16].

Through sensor fusion, the robot can recognize dynamic changes around the user's head in real-time and adjust its path based on feedback data. For example, when LiDAR detects that the user's ear is approaching the tools, the control system will immediately adjust the manipulator's trajectory to avoid contact. Meanwhile, the visual system monitors the smoothness of the path, ensuring continuity and stability.

Sliding Mode Control is a nonlinear control technique widely used in robot obstacle avoidance and path correction. It processes the system's motion state in segments and applies different control strategies in different states to achieve precise control in complex environments. In haircutting robots, sliding mode control ensures that the manipulator maintains the smoothness and stability of the trajectory while avoiding obstacles.

The advantage of sliding mode control lies in its robustness against system uncertainties and external disturbances. Even when the user suddenly moves or the tools deviate from the trajectory, the sliding mode control system can quickly correct the path, ensuring the continuity and safety of the operation [17].

Neural network adaptive control technology provides a higher level of intelligent control capability for haircutting robots. Through adaptive learning, the robot can automatically adjust its operating parameters based on real-time data and environmental changes, achieving more flexible and precise control.

The core of the neural network adaptive control system is its online learning capability. By continuously acquiring real-time data during the haircutting process, the neural network can automatically adjust control strategies, optimizing the cutting path and force. This online learning technology enables the robot to make personalized adjustments based on the user's individual characteristics (such as hair length, thickness, and texture), ensuring efficiency and comfort [18].

Neural network adaptive control can also perform self-adjustment in complex environments. For example, when the user's head shape or haircutting requirements change, the neural network can learn and adapt to these changes, automatically generating new cutting paths and operation strategies. This flexible adjustment capability greatly enhances the intelligence level of the haircutting robot, enabling it to handle diverse scenarios and requirements.

References

1. Sciavicco, L., & Siciliano, B. (2012). *Modelling and control of robot manipulators*. Springer Science & Business Media.
2. Nakamura, Y., Hanafusa, H., & Yoshikawa, T. (1987). Task-priority based redundancy control of robot manipulators. *The International Journal of Robotics Research, 6*(2), 3–15.
3. Williams, R. J., & Zipser, D. (1995). Gradient-based learning algorithms for recurrent networks and their computational complexity. In *Backpropagation: Theory, architectures, and applications* (pp. 433–486). Lawrence Erlbaum Associates.
4. Zhang, Y., Li, S., & Liu, X. (2018). Neural network-based model-free adaptive near-optimal tracking control for a class of nonlinear systems. *IEEE Transactions on Neural Networks and Learning Systems, 29*(12), 6227–6241.
5. Zadeh, L. A. (1996). *Fuzzy sets, fuzzy logic, fuzzy systems: Selected papers*. World Scientific.
6. Holland, J. H. (1992). *Adaptation in natural and artificial systems*. MIT Press.
7. Hart, P. E., Nilsson, N. J., & Raphael, B. (1968). A formal basis for the heuristic determination of minimum cost paths. *IEEE Transactions on Systems Science and Cybernetics, 4*(2), 100–107.
8. Dijkstra, E. W. (1959). A note on two problems in connexion with graphs. *Numerische Mathematik, 1*(1), 269–271.
9. LaValle, S. M. (1998). *Rapidly-exploring random trees: A new tool for path planning* (Technical Report).
10. Bellman, R. (1957). *Dynamic programming*. Princeton University Press.
11. De Boor, C. (1978). *A practical guide to splines*. Springer-Verlag.
12. LeCun, Y., Bengio, Y., & Hinton, G. (2015). Deep learning. *Nature, 521*(7553), 436–444.

13. Hogan, N. (1985). Impedance control: An approach to manipulation: Part I—Theory. *Journal of Dynamic Systems, Measurement, and Control, 107*(1), 1–7.
14. Åström, K. J., & Hägglund, T. (1995). *PID controllers: Theory, design, and tuning*. Instrument Society of America.
15. Zhang, Z. (2012). Microsoft Kinect sensor and its effect. *IEEE Multimedia, 19*(2), 4–10.
16. Borenstein, J., & Koren, Y. (1988). Real-time obstacle avoidance for fast mobile robots in cluttered environments. In *IEEE international conference on robotics and automation.*
17. Utkin, V. I. (1992). *Sliding modes in control and optimization.* Springer-Verlag.
18. Werbos, P. J. (1990). Backpropagation through time: What it does and how to do it. *Proceedings of the IEEE, 78*(10), 1550–1560.

Human–Computer Interaction and User Experience

12

1 User Interface Design and Interaction Methods

In the field of human–computer interaction, user interface design is crucial in determining user experience, especially when users lack a technical background. Designing a simple and intuitive interface that allows hairstylists to easily operate haircutting robots has become a focal point [1]. Traditional robot operations often require programming skills, which is unrealistic for most hairstylists. Therefore, simplifying the interaction interface is essential for enhancing user experience.

1.1 Simplified User Interface Design

To enable hairstylists to efficiently operate the robot without programming knowledge, the user interface needs to be as simplified as possible. An ideal interface could provide visual operation methods, allowing hairstylists to easily plan haircutting paths by dragging icons or selecting preset hairstyle templates. Such a design not only saves time but also reduces the cost for hairstylists to learn new technologies [2] (Fig. 1).

Additionally, the haircutting robot should have preset modes where the hairstylist only needs to select an appropriate mode to start operation. These modes can be automatically recommended and adjusted based on different hairstyle requirements, user facial features, and hair types. For the user, the operation becomes less cumbersome, and the robot automatically generates the corresponding operational procedures, requiring only simple confirmation from the hairstylist.

© The Author(s), under exclusive license to Springer Nature Switzerland AG 2025

S. Li, *Haircutting Robots*, https://doi.org/10.1007/978-3-031-84026-5_12

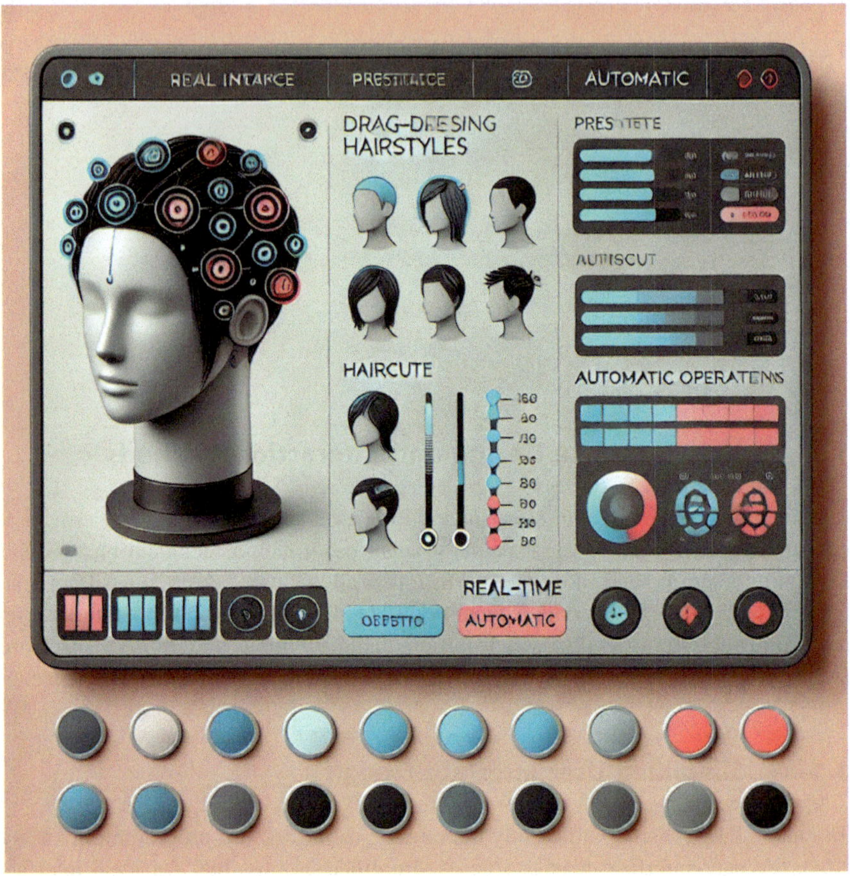

Fig. 1 A simple and intuitive user interface is crucial for the widespread adoption of haircutting robots

1.2 Natural Language Interface Empowered by Large Models

The introduction of large models has greatly enhanced the intelligent interaction capabilities of robot systems, especially in natural language processing [3]. Hairstylists can directly converse with the robot via voice or text, such as "design a short hairstyle for the customer" or "trim the bangs" [4]. The language processing system powered by large models parses these complex natural language requirements and automatically generates haircutting paths and operational steps. Through this natural conversational interaction, hairstylists can easily operate the robot without any programming knowledge.

In practical applications, large models can understand not only simple commands but also complex requirements. For example, a hairstylist can tell the robot,

"Design a hairstyle suitable for a round face for the customer, and ensure the bangs look natural." The large model analyzes these requirements, generates the most suitable hairstyle design plan, and automatically plans detailed haircutting steps, greatly improving operational convenience and efficiency.

1.3 Visual Feedback and Real-Time Adjustment Interface

Real-time visual feedback is also crucial in user interface design. By integrating depth sensors and 3D modeling technology, hairstylists can view the robot's haircutting progress and the final hairstyle effect on the screen in real-time. This visual interface not only helps hairstylists monitor the haircutting process at any time but also allows immediate adjustment of the haircutting path, ensuring that the hairstyle design meets user expectations [5].

During the haircutting process, the robot interface provides multiple views for the hairstylist to view different angles of the hairstyle in real-time. These views help hairstylists precisely adjust hairstyle details, such as the length of edge trimming and the treatment of layers. Additionally, an intelligent assistant based on large models can provide real-time suggestions according to the haircutting progress, prompting the hairstylist whether further trimming or hairstyle adjustments are needed.

The integration of large models makes this real-time adjustment mechanism more intelligent. The system can not only understand the hairstylist's adjustment instructions but also proactively provide suggestions, such as proposing appropriate hairstyle trimming adjustments based on the user's face shape, hair density, and color. The combination of visual feedback and real-time adjustment enhances the precision of haircutting, ensuring that the results are more perfect.

This design not only reduces operational complexity but also improves the hairstylist's work efficiency. By combining the language understanding capabilities of large models with real-time feedback, hairstylists can easily collaborate with robots to complete complex haircutting tasks without relying on technical support.

2 Personalized Hairstyle Customization and Recommendation System

In modern haircutting robots, personalized customization and hairstyle recommendation systems are important technical means to enhance user experience. By combining big data, machine learning, and augmented reality (AR) technologies, haircutting robots can not only provide personalized hairstyle recommendations but also display virtual hairstyle effects to users, helping them make the best choice (Fig. 2).

Fig. 2 By using recommendation system technology in AI systems, even when customers cannot clearly express their desired hairstyle, the system can accurately capture customer preferences and find the hairstyle they like by analyzing the time the customer stays on each image while browsing the hairstyle library

2.1 Recommendation Algorithms Based on Big Data

The core of the hairstyle recommendation system lies in its capabilities based on big data and machine learning technologies. By collecting users' historical hairstyle data, browsing habits, and personal facial features, the haircutting robot can utilize recommendation algorithms to generate personalized hairstyle design schemes [6]. This data may include the hairstyles previously chosen by the user, preferences for specific hairstyle styles, browsing history, and evaluations of certain hairstyles.

The recommendation system constructs a user's hairstyle preference model by analyzing this data. For example, if a user prefers short hair, curly hair, or layered hairstyles, the system analyzes these features and generates a recommendation list. On this basis, the haircutting robot, combined with the powerful predictive capabilities of large models, can recommend the most suitable hairstyle scheme based on the user's facial shape, hair texture, and personal style. This personalized recommendation not only enhances user satisfaction but also greatly simplifies the process of selecting hairstyles.

Moreover, the recommendation algorithm continuously optimizes with user usage. Each time the user selects a hairstyle from the recommendation list, the system updates the user's preference model. Through continuous learning and optimization, the recommendation system can increasingly accurately predict the hairstyle styles the user may like.

2.2 Hairstyle Style Classification and User Preference Analysis

Users' hairstyle preferences are usually diverse, but they are not always clear about what hairstyle suits them best. The haircutting robot can analyze users' selection behaviors through large models to help them find potential favorite hairstyles. The large model can extract potential preference patterns from the user's browsing history, and the system infers user preferences based on these patterns. For example, some users may tend to choose shorter hairstyles, while others prefer hairstyles with layers.

By classifying these preferences, the recommendation system can push hairstyles in a targeted manner [4, 7]. For instance, after a user has browsed several hairstyles with natural curls, the system automatically pushes more hairstyles with similar features. The advantage of large models is that they can handle complex multidimensional data, comparing factors such as the user's facial shape, hair texture, and color with samples in the hairstyle library to generate accurate recommendation lists.

Additionally, the system can classify based on different hairstyle features, such as length, curliness, hair color, etc. Through this classification, users can quickly find the hairstyle styles they like, and the system continuously optimizes recommendation accuracy based on user choices. Ultimately, users can find hairstyles that match their style in a short time through the recommendation system [8], reducing tentative hairstyle selections and improving satisfaction.

2.3 Virtual Hairstyle Try-On and Augmented Reality Technology (AR)

With the development of image generation and augmented reality (AR) technologies, virtual try-on has become an important function in personalized hairstyle recommendations. Users can preview different hairstyle effects in real-time through

AR technology without actually cutting their hair [9]. Through cameras, the haircutting robot can generate a 3D model of the user's head and accurately overlay different hairstyles onto the user's head.

The application of large models in image generation can enhance the realism of hairstyle try-on. The system can generate effects of different hairstyles based on the user's hair length, color, and density, and can also adjust hairstyle details according to the user's facial expressions, making the hairstyle try-on effect closer to reality. For example, the system can dynamically adjust the length of the bangs or recommend suitable hairstyle angles based on the user's face shape.

Virtual hairstyle try-on provides users with an immersive experience through augmented reality technology. Users can see the effects of different hairstyles on themselves in real-time, allowing them to make more informed decisions without changing their existing hairstyle. This not only reduces the risk of haircutting but also increases users' confidence in hairstyle design.

By combining big data, recommendation algorithms, and augmented reality technology, haircutting robots can provide a more personalized hairstyle design experience. The application of large models enables the system to better understand user needs, automatically generate personalized hairstyle recommendations, and help users find their ideal hairstyle through virtual hairstyle try-on.

3 Voice Control and Gesture Recognition Technology

In modern human–computer interaction technologies, voice control and gesture recognition provide robots with more natural operation methods. Empowered by large models, haircutting robots can accurately understand users' voice commands and respond to hairstylists' gestures in real-time, making the operation process more concise and smooth. The combination of voice control and gesture recognition will greatly enhance the flexibility and intelligence level of haircutting robots.

3.1 Voice Recognition System Driven by Large Models

Voice control technology is an important development direction for future interaction with haircutting robots, especially through the voice recognition capabilities of large models. The system can understand the hairstylist's natural language instructions and respond in real-time. Large models provide extremely high recognition accuracy and flexibility in multilingual and diverse operating environments by processing complex language structures and contexts.

In practical applications, hairstylists only need to control the haircutting robot through simple voice commands [3, 7]. For example, a hairstylist can issue commands like "trim the bangs" or "cut two centimeters shorter," and the system automatically parses these instructions and converts them into specific operation plans [9]. By deeply understanding natural language, large models can handle not only simple commands but also parse complex semantics. For instance, a hairstylist can tell the robot: "Design a short hairstyle suitable for a round face for the customer, maintaining a natural transition on the sides." The large model can understand such complex requirements and, combined with the user's facial features and hairstyle styles, automatically plan the haircutting path.

The biggest advantage of voice control lies in its naturalness and convenience. Hairstylists do not need to use manual control panels or learn complex operating steps; they only need to issue voice commands, and the robot can efficiently execute tasks. Moreover, voice control, combined with the deep learning technology of large models, can continuously optimize and improve recognition accuracy. The system learns from the user's command data, gradually adapting to individual voice characteristics and operating habits, thereby enhancing the interaction experience.

3.2 Gesture Recognition Technology

Gesture recognition technology captures the hairstylist's hand movements in real-time through cameras and depth sensors, providing a non-contact control method. Similar to voice control, gesture recognition allows hairstylists to interact more flexibly with the robot and achieve precise control over the haircutting process. The introduction of large models enables the gesture recognition system to more accurately analyze and understand complex gestures, significantly improving recognition accuracy [8].

During the haircutting process, hairstylists can control the robot through simple gestures. For example, a hairstylist can start or pause the robot with a "wave" gesture or adjust the haircutting area and intensity through different gestures. The gesture recognition system captures these actions in real-time and, based on calculations from the large model, determines the meaning of the gestures and issues corresponding operation commands.

Another important application of gesture recognition is the enhancement of automation in "multi-step operations." By detecting continuous gestures, hairstylists can design multi-step haircutting processes, such as from "trimming the neckline" to "thinning the hair ends." The large model can recognize these continuous gestures and automatically connect the corresponding operations, reducing manual operation steps for the hairstylist.

3.3 Precise Recognition and Real-Time Adjustment Empowered by Large Models

Large models play a critical role in voice and gesture recognition, enhancing the intelligence and flexibility of these technologies. Traditional voice and gesture recognition systems often can only handle simple, fixed commands or gestures, whereas large models can process more complex and natural interactions through deep learning capabilities [10]. For example, large models can adjust the hairstylist's instructions according to different operating scenarios. In a busy salon environment, the system can automatically filter out irrelevant background noise or actions through contextual understanding, capturing only key commands or gestures.

Moreover, each action and instruction of the hairstylist can be learned and adapted in real-time through the large model. The system continuously optimizes recognition accuracy, adjusting recognition algorithms based on each hairstylist's operating habits, providing a personalized operating experience for each user. For instance, some hairstylists may prefer using more subtle gestures to command the robot; the system can adjust according to these differentiated features to ensure recognition accuracy and response speed.

3.4 Combination of Gesture and Voice: Enhancing Multimodal Interaction Experience

The combination of voice control and gesture recognition technologies will further enhance the interaction experience of haircutting robots. Hairstylists can issue general operation commands through voice and make precise detail adjustments through gestures. For example, a hairstylist can tell the robot "start haircutting" while using gestures to precisely control the robot's movement position, angle, and cutting force on the customer's head. This combination of multimodal interaction greatly increases the robot's flexibility and response speed, making the haircutting process smoother and more efficient.

The combined interaction of voice and gestures also significantly enhances the operability of the haircutting robot, especially when hairstylists need to manage multiple tasks simultaneously. The combination allows hairstylists to control the robot and other haircutting tools at the same time, making operations more diversified and efficient.

Empowered by large models, voice control and gesture recognition technologies not only make haircutting robots more intelligent but also make human–computer interaction more natural and smooth. Large models enhance recognition accuracy and optimize the operation experience through continuous learning, providing hairstylists and users with more convenient intelligent haircutting services.

4 Facial Expression Recognition and User Satisfaction Feedback

With the continuous development of artificial intelligence and computer vision technology, haircutting robots can not only perform precise haircutting tasks but also understand the customer's emotional state in real-time through facial expression recognition technology. Empowered by large models, robots can better understand user feedback and make corresponding adjustments, thereby improving overall service satisfaction.

4.1 Real-Time Facial Expression Recognition Technology

Facial expression recognition technology is a direct means of obtaining emotional feedback from users. During the haircutting process, users' facial expressions often reflect their attitudes toward the current hairstyle design and the haircutting process. By integrating cameras and computer vision systems, the haircutting robot can recognize users' facial expressions in real-time, such as happiness, nervousness, or dissatisfaction [10]. This emotional feedback is crucial for adjusting services. For example, when the system detects that a user is showing signs of dissatisfaction, the haircutting robot can pause the current haircutting operation and prompt the hairstylist to reconfirm the hairstyle design with the user, avoiding unnecessary mistakes [11].

The introduction of large models makes facial expression recognition technology more intelligent, capable of handling complex facial features and diverse expressions. The system can not only recognize common emotional states but also make more precise emotional judgments based on subtle facial changes. For instance, when a user slightly frowns, the system can recognize that the user may be dissatisfied with the current hairstyle design and automatically suggest that the hairstylist make adjustments. Through this precise feedback mechanism, the haircutting robot can effectively enhance the user's haircutting experience (Fig. 3).

4.2 Customer Experience Analysis and Improvement

Facial expression recognition can not only provide real-time feedback on the user's emotional state but also use accumulated emotional data over time to optimize the haircutting robot's operational processes and service quality. By recording users' facial expression changes during each haircut, the system can learn their emotional patterns and integrate this data into subsequent hairstyle design and recommendation systems [12].

For example, when the system identifies that certain hairstyle designs typically elicit positive emotions from users, it can prioritize these designs in the hairstyle recommendation system, thereby improving user satisfaction. Conversely, if certain hairstyles or haircutting steps repeatedly trigger negative emotions, the system will automatically avoid these designs in future operations or prompt the hairstylist to adjust the design before the haircut [13].

This learning mechanism based on emotional feedback can significantly enhance the haircutting robot's level of personalized service, ensuring that each haircut not only meets the user's aesthetic needs but also keeps the user in a

Fig. 3 Facial expression recognition technology can accurately capture customer satisfaction with the hairstyle

pleasant mood throughout the process. Moreover, by comparing emotional feedback from different users, the system can analyze common factors affecting user satisfaction and optimize the overall haircutting process accordingly [14].

By analyzing and improving based on user emotional data, the haircutting robot can continuously optimize its operating modes, enhance user experience, and thereby establish more trusting and enjoyable customer relationships.

5 How Large Models Empower Haircutting Robots

With the continuous development of artificial intelligence technology, large language models (LLMs) have made significant progress not only in natural language processing but also in various robotic applications. In the field of haircutting

Fig. 4 Large models empower haircutting robots through natural language processing, feedback learning, and personalized recommendations

robots, the introduction of large models offers new possibilities in intelligent interaction, hairstyle design, real-time adjustment, and adaptive learning (Fig. 4).

5.1 Natural Language Processing and Intelligent Interaction

The natural language processing capabilities of large models enable haircutting robots to understand and parse complex language commands, enhancing the interaction experience. In traditional robot operations, hairstylists often need to input commands through programming or preset control panels, which is a significant barrier for those without a technical background. Empowered by large models, hairstylists can simply control the robot using natural language commands, such as "trim the bangs" or "cut two centimeters shorter" [15].

Large models can handle not only simple commands but also complex language scenarios. For example, a hairstylist can tell the robot, "Design a short hairstyle suitable for a long face and maintain distinct layers at the ends." The large model can parse these complex requirements and automatically convert them into specific technical execution steps, such as hairstyle design, path planning, and haircutting action generation [16]. This process significantly lowers the technical operation threshold, enabling more hairstylists to easily use intelligent haircutting robots.

5.2 Deep Learning and Hairstyle Generation

By combining deep learning technology, large models can learn various hairstyle design styles from vast amounts of hairstyle data. By analyzing the user's facial shape, hair texture, personal style, and past hairstyle preferences, the system can generate personalized hairstyle design schemes in real-time. Large models possess powerful data processing capabilities, enabling them to generate the most suitable hairstyle designs based on the user's unique facial features and display the hairstyle effects through real-time simulation [17].

This deep learning–based hairstyle generation greatly enhances design accuracy. Previous hairstyle recommendations might rely on simple geometry or template matching, but large models combined with user data can generate designs that better match personal styles. This not only improves user satisfaction but also reduces the time needed for hairstyle recommendations. Through rapid processing and computation, hairstylists can customize hairstyles for each customer and confirm the design with them via virtual simulation before cutting, avoiding unnecessary modifications [18].

5.3 Real-Time Adjustment and Learning Ability

Another major advantage empowered by large models is their adaptive learning capability. By continuously analyzing each customer's feedback data, the haircutting robot can gradually optimize its decision-making processes. During the haircutting process, the robot can dynamically adjust the haircutting plan based on the user's immediate feedback. For example, if the user shows dissatisfaction or suggests modifying certain parts during the haircut, the robot can adjust the haircutting path or hairstyle design in real-time [19].

This adaptive learning is not limited to a single haircut but can accumulate experience over long-term use. Data from each haircut is stored and analyzed, allowing the robot to extract valuable patterns to optimize future operations. For instance, if a user prefers certain elements in a hairstyle (like layering or curliness), the robot will remember these preferences and prioritize these features in the next haircut. This self-learning ability driven by large models ensures that the haircutting robot can, over time, provide increasingly refined services that meet customer expectations [20].

Large models not only endow the haircutting robot with powerful intelligent capabilities but also greatly improve operational efficiency and flexibility. Through continuous learning and optimization, the robot can provide personalized, high-precision haircutting services for each customer, thereby enhancing the overall user experience.

References

1. Goodfellow, I., Bengio, Y., & Courville, A. (2016). *Deep learning*. MIT Press.
2. Jurafsky, D., & Martin, J. H. (2020). *Speech and language processing*. Pearson.
3. Krizhevsky, A., Sutskever, I., & Hinton, G. E. (2012). ImageNet classification with deep convolutional neural networks. *Advances in Neural Information Processing Systems, 25*, 1097–1105.
4. He, X., Liao, L., Zhang, H., Nie, L., Hu, X., & Chua, T.-S. (2017). Neural collaborative filtering. In *Proceedings of the 26th international conference on world wide web* (pp. 173–182).
5. Zhang, Y., & Yang, Q. (2020). A survey on multi-task learning. *IEEE Transactions on Knowledge and Data Engineering, 34*(12), 5586–5609.
6. Fu, M., Qu, H., Yi, Z., Lu, L., & Liu, Y. (2018). A novel deep learning-based collaborative filtering model for recommendation system. *IEEE Transactions on Cybernetics, 49*(3), 1084–1096.
7. Graves, A., Mohamed, A.-R., & Hinton, G. (2013). Speech recognition with deep recurrent neural networks. In: *IEEE international conference on acoustics, speech and signal processing* (pp. 6645–6649).
8. Shotton, J., Sharp, T., Kipman, A., Fitzgibbon, A., Finocchio, M., Blake, A., Cook, M., & Moore, R. (2013). Real-time human pose recognition in parts from single depth images. *Communications of the ACM, 56*(1), 116–124.
9. Mikolov, T., Karafiát, M., Burget, L., Cernocký, J., & Khudanpur, S. (2010). Recurrent neural network based language model. *Interspeech, 2*(3), 1045–1048.
10. Wang, M., & Deng, W. (2021). Deep face recognition: A survey. *Neurocomputing, 429*, 215–244.
11. Lopes, A. T., de Aguiar, E., De Souza, A. F., & Oliveira-Santos, T. (2017). Facial expression recognition with convolutional neural networks: Coping with few data and the training sample order. *Pattern Recognition, 61*, 610–628.
12. Dhall, A., Goecke, R., Lucey, S., & Gedeon, T. (2012). Collecting large, richly annotated facial-expression databases from movies. *IEEE Multimedia, 19*(3), 34–41.
13. Liu, M., Li, S., Shan, S., & Chen, X. (2014). AU-aware deep networks for facial expression recognition. In *IEEE international conference on automatic face and gesture recognition*, pp. 1–6.
14. Cowie, R., McKeown, G., & Douglas-Cowie, E. (2012). Tracing emotion: An overview. *International Journal of Synthetic Emotions, 3*(1), 1–17.
15. Vaswani, A., Shazeer, N., Parmar, N., Uszkoreit, J., Jones, L., Gomez, A. N., Kaise, L., & Polosukhin, I. (2017). Attention is all you need. *Advances in Neural Information Processing Systems, 30*, 5998–6008.
16. Devlin, J., Chang, M. W., Lee, K., & Toutanova, K. (2019). BERT: Pre-training of deep bidirectional transformers for language understanding. In *Proceedings of the 2019 conference of the North American chapter of the association for computational linguistics: Human language technologies* (pp. 4171–4186).
17. Radford, A., Wu, J., Child, R., Luan, D., Amodei, D., & Sutskever, I. (2019). *Language models are unsupervised multitask learners*. OpenAI.

18. Dosovitskiy, A., Beyer, L., Kolesnikov, A., Weissenborn, D., Zhai, X., Unterthiner, T., Dehghani, M., Minderer, M., Heigold, G., Gelly, S., Uszkoreit, J., & Houlsby, N. (2021). An image is worth 16x16 words: Transformers for image recognition at scale. In *Proceedings of the 9th international conference on learning representations.*
19. Brown, T., Mann, B., Ryder, N., Subbiah, M., Kaplan, J. D., Dhariwal, P., Neelakantan, A., Shyam, P., Sastry, G., Askell, A., & Agarwal, S. (2020). Language models are few-shot learners. *Advances in Neural Information Processing Systems, 33*, 1877–1901.
20. Ramesh, A., Pavlov, M., Goh, G., Gray, S., Voss, C., Radford, A., Chen, M., & Sutskever, I. (2021). Zero-shot text-to-image generation. In *Proceedings of the 38th international conference on machine learning* (pp. 8821–8831).

Safety and Reliability Design 13

1 Safety Risk Analysis During the Haircutting Process

As an automated device, haircutting robots come into direct contact with the human body during the haircutting process, making their safety particularly important. Without adequate safety protection measures, their use can not only result in property loss but also pose a threat to the personal safety of customers. Therefore, the safety design of haircutting robots must comprehensively cover all potential risk points. Below are common types of safety issues and their risk analyses encountered during the use of haircutting robots.

Electrical Leakage:

Electrical safety issues are a major risk faced by haircutting robots, especially when operating in environments exposed to water or high humidity. Haircutting robots typically integrate numerous electrical components, which, without proper insulation and protective measures, may lead to electrical leakage due to wiring aging or electrical failures, posing an electric shock hazard to both customers and operators. This risk must be mitigated through the use of residual current devices (RCDs) and insulating materials in electrical design. Additionally, regular inspections and maintenance of the electrical systems are essential to prevent such occurrences [1] (Fig. 1).

Mechanical Arm Injuries:

The manipulator arm of a haircutting robot is the core component that comes into direct contact with the customer's head. Therefore, insufficient control precision or sensor failures can lead to injuries. During operation, if the manipulator fails to accurately sense the position or force applied to the customer's head, excessive contact force may be exerted, resulting in skin or scalp damage. Furthermore, the high speed and complex motion trajectories of the manipulator during haircutting increase the risk of collisions or misoperations [2]. To address this, the safety

© The Author(s), under exclusive license to Springer Nature Switzerland AG 2025

S. Li, *Haircutting Robots*, https://doi.org/10.1007/978-3-031-84026-5_13

Fig. 1 Haircutting robots must consider a range of potential safety risks, including electrical insulation and excessive force that could harm customers

design of the manipulator should include force sensors and speed limitations to ensure precise and safe movements.

Clipping Hair or Cutting Skin:

The haircutting tools used by the robot are another significant safety concern. Improper operation of the haircutting tools can result in hair being pinched, causing pain or pulling, or if the tools come too close to the skin, they might inadvertently cut the skin, leading to injuries [3]. Utilizing high-precision distance sensors and intelligent haircutting algorithms can effectively reduce the occurrence of such risks.

Sudden Power Outages:

Sudden power outages during the haircutting process can leave the robot in an abnormal state, potentially causing safety issues. Especially when the manipulator arm or haircutting tools are in operation, an unexpected power loss may cause the tools to get stuck in the customer's hair or fail to retract promptly, leading to discomfort or potential injury. Therefore, haircutting robots need to incorporate emergency power supplies or mechanisms that allow for safe shutdown in the event of a power outage, ensuring that all ongoing operations are handled smoothly [4].

Network Interruptions:

With the widespread adoption of Internet of Things (IoT) and cloud computing technologies, many haircutting robots rely on network connectivity for remote control, data transmission, and real-time updates. If the network suddenly disconnects, the robot may be unable to continue operations or may experience program freezes and stalls due to data transmission failures. The safety hazards posed by network interruptions cannot be overlooked [5]. Systems should be equipped with local processing capabilities to continue executing local tasks or safely shut down when external network connections are unavailable (Fig. 2).

Program malfunctions:

Program malfunctions or software failures are another potential source of severe safety issues. Haircutting robots rely on complex software control systems, and any code errors or system crashes can result in the manipulator arm or haircutting tools becoming uncontrollable, executing incorrect actions that endanger customer safety [6]. To prevent such issues, rigorous program testing and validation should be conducted during the development process. Additionally, incorporating redundancy mechanisms and fault recovery systems into the system design ensures that safe operations are maintained even in the event of program failures (Fig. 3).

Hacker Attacks:

Fig. 2 Program malfunctions can cause haircutting robots to operate outside their intended settings

Fig. 3 Network security for haircutting robots includes preventing hacker attacks that could control the robot to harm customers or tamper with customized hairstyles

As haircutting robots become increasingly networked, the risks associated with cybersecurity also rise. If a haircutting robot system is compromised by hackers, malicious software could control the robot to perform dangerous operations or even steal customers' personal privacy information [7]. Therefore, network security protections for haircutting robots should include data encryption, access controls, and multi-layered firewall protections to safeguard against hacker attacks.

Network security measures should not only protect customer personal data but also prevent malicious software from controlling the robot to perform hazardous actions. The system can monitor network traffic in real-time to identify and block abnormal access requests while ensuring the security of all data transmissions. Network security protection should be a core part of system design, with regular updates and vulnerability patches to prevent system intrusions [8].

Safety of Haircutting Equipment Materials:

The safety of materials used in haircutting tools, manipulator arm housings, and other components is also a critical consideration in design. Using inappropriate materials may lead to skin allergies or structural failures due to material degradation. Equipment materials should comply with national and international standards, ensuring they are non-toxic, harmless, and durable for long-term use [8].

Sudden Customer Health Issues:

During the haircutting process, if a customer suddenly experiences health issues such as dizziness or nausea, it could impact the safety of the operation. Haircutting robots should have the capability to monitor the customer's physical state in real-time. When abnormal conditions are detected, the robot should immediately stop operations and alert the hairstylist or the customer to intervene. This not only helps in reducing the occurrence of accidents but also ensures that the robot can respond swiftly to any emergencies [9].

Through a detailed analysis of these common safety issues, a better understanding of the risk points in the design of haircutting robots can be achieved. Appropriate protective measures can then be implemented in subsequent safety designs to ensure the safety and reliability of operations.

2 Multi-level Safety Protection Mechanisms

To ensure the safe operation of haircutting robots, system design must incorporate multi-level safety protection mechanisms. These mechanisms include comprehensive protective measures from hardware to software, aiming to minimize any potential safety risks to the greatest extent. Under a multi-level protection framework, different types of safety risks can be effectively monitored and managed.

2.1 Design of Safety Check Layer

Haircutting robots need to establish a high-priority safety check layer through sensors and real-time monitoring systems. When sensors detect any potential safety hazards, such as collisions, excessive contact force, or equipment malfunctions, the system should immediately halt all operations and enter a safe mode. This mechanism is designed to prevent injuries at the earliest moment, ensuring the safety of customers [10].

The safety check layer can integrate various sensor technologies, including force feedback sensors, infrared distance sensors, and visual sensors, to monitor the robot's operational status in real-time. When abnormal sensor data is detected, the system will prioritize stopping the haircutting actions and prompt the hairstylist to perform a manual inspection [11].

2.2 Electrical Safety and Leakage Protection

Electrical safety is central to the safety design of haircutting robots. Since haircutting devices may come into contact with water sources or operate in humid environments, robust leakage protection is essential. Haircutting robots should be equipped with highly sensitive residual current devices (RCDs). When the system detects electrical leakage, it should immediately cut off the power supply to prevent electric shock accidents [12]. Additionally, the entire device housing should use insulating materials and ensure that internal electrical components have good waterproof and moisture-resistant properties.

Electrical safety design should extend beyond the device itself to include external power interfaces and cables. For example, the robot can detect the condition of cables and promptly report any signs of aging or damage to prevent leakage incidents.

2.3 Mechanical Safety Protection

The manipulator arm and haircutting tools of a haircutting robot are parts that come into direct contact with customers, making mechanical safety protection crucial. Force feedback sensors can monitor the force between the manipulator arm and the customer in real-time, ensuring that the robot does not apply excessive pressure that could cause skin damage [3]. This force feedback system can adaptively adjust based on different customer needs, ensuring both safety and comfort.

Moreover, haircutting robots can implement speed limitations. When the manipulator arm or haircutting tools approach the customer's head, the system can automatically reduce speed to prevent accidental injuries caused by high-speed movements. Combined with intelligent algorithms, the system can also predict the manipulator arm's motion trajectory, proactively avoiding potential collisions or misoperations [1].

In the event of emergencies such as power outages, program malfunctions, or hardware failures, haircutting robots must have emergency stop mechanisms. Emergency stop mechanisms allow the system to immediately halt all mechanical operations upon detecting anomalies, preventing further injuries [2]. Additionally, haircutting robots can be designed with redundant systems to ensure that in the event of failures in critical components, the system can quickly switch to backup systems to continue safely shutting down operations.

To ensure the reliability of emergency stop mechanisms, the system should regularly conduct simulation drills to test response times and accuracy under various extreme conditions. Redundant designs can also include power redundancies, ensuring that even in sudden power losses, the system has sufficient power to complete safe shutdown operations [4].

As haircutting robots become networked, cybersecurity becomes a significant concern. To prevent hacker attacks, systems must implement multi-layered network security protection measures, including data encryption, identity authentication, and firewall technologies [7]. Network security protection not only involves safeguarding customers' personal data privacy but also preventing malicious software from controlling the robot to perform dangerous operations.

The system can monitor network traffic in real-time to identify and block abnormal access requests while ensuring the security of all data transmissions. Network security protection should be a core component of system design, with regular updates and vulnerability patches to prevent system intrusions [8].

Haircutting robots also need to have real-time monitoring functions for environmental conditions and customer health status. By using sensors during operations to detect customers' physiological signals such as body temperature and breathing rate, the system can promptly identify abnormal reactions like dizziness or fainting [13]. Upon detecting customer discomfort, the system can automatically stop haircutting operations and alert the hairstylist or the customer to intervene. This not only effectively reduces the occurrence of accidents but also ensures swift responses to sudden customer health issues.

Environmental monitoring systems can also track the salon's humidity, temperature, and air quality to ensure comfort and safety during haircutting. For example, when high humidity levels are detected, the system can automatically adjust the device's operational parameters to prevent the impact of a humid environment on the equipment and operation process [14].

3 System Reliability and Stability Testing

As a complex automated device, the reliability and stability of haircutting robots are crucial for ensuring safe and continuous operations. This section discusses how to enhance the system's reliability and stability through various tests and preventive measures.

The mechanical structure's stability directly affects the long-term operation and safety of haircutting robots. To assess the durability of mechanical components, haircutting robots should undergo periodic fatigue testing and lifespan prediction. These tests typically simulate the actual usage conditions of the manipulator arm and haircutting tools, detecting fatigue cracks and deformation in materials to predict the lifespan of mechanical parts. Through this approach, potential problematic components can be identified and replaced in a timely manner, ensuring the robot's stability and safety during long-term operation [15].

To reduce failure rates and maintenance costs, haircutting robots should be equipped with predictive maintenance systems. These systems use sensors installed on critical components to monitor the working status and performance indicators of mechanical parts in real-time. By utilizing data analysis and machine learning algorithms, the system can identify early signs of potential failures, such as abnormal temperature, pressure changes, or vibration frequency variations, and issue early warnings for maintenance actions. Predictive maintenance not only reduces the occurrence of unexpected failures but also extends the lifespan of mechanical components, enhancing the device's reliability [16].

Vibration inspection is an essential method for identifying early signs of mechanical faults. Haircutting robots can monitor the vibration signals of the manipulator arm and haircutting tools through vibration sensors installed on critical components. Abnormal vibration patterns may indicate issues such as loosened parts, bearing wear, or imbalance problems. If not addressed promptly, these issues could lead to equipment damage or safety hazards. By analyzing vibration signals, the system can accurately diagnose and locate the fault causes, improving the accuracy and efficiency of fault diagnosis [12] (Fig. 4).

To ensure the stability of haircutting robots under different environmental conditions, multi-level stability testing is necessary. These tests include assessing the impact of environmental factors such as temperature, humidity, and lighting on the device's operation, as well as evaluating performance under long-term use. By

Fig. 4 Haircutting robots may be deployed in various environments such as subway waiting areas, shopping malls, and outdoor parks, necessitating comprehensive reliability and stability testing for different scenarios

simulating various working conditions in both laboratory and real-world scenarios, the robot's adaptability and stability in different environments can be assessed, allowing for design and manufacturing process optimizations to enhance overall performance and reliability [3].

Key components of haircutting robots should be designed with redundancy to handle potential failures and system malfunctions. For example, in the event of failures in critical components such as power supplies, sensors, or actuators, the system should have the capability to switch to backup systems to ensure continuous and safe operations. Redundant designs can also include backup power supplies, dual control systems, or backup communication networks to handle unexpected situations and emergencies [1].

By integrating the above various tests and safety measures, haircutting robots can maintain stable operations under various complex environments and long-term usage conditions, while maximizing operational safety and device reliability.

4 Fault Analysis and Improvement Mechanism

Fault analysis and improvement mechanisms are key aspects to ensure the continuous reliability and efficiency of haircutting robots. By thoroughly analyzing the causes of faults and employing data-driven decision-making processes, the performance and safety of the robot can be significantly enhanced.

Fault analysis of haircutting robots involves not only identifying the immediate causes of faults but also delving into their root causes, which typically encompass aspects such as system design, operational procedures, and maintenance strategies [17]. By collecting and analyzing fault data, including operation logs, sensor data, user feedback, and historical maintenance records, technical teams can utilize statistical analysis and machine learning methods to identify fault patterns and potential risk points from the data.

Once the fault causes are determined, the haircutting robot system can enter an optimization phase. This phase primarily involves developing and implementing targeted improvement measures, such as adjusting the robot's design, updating software algorithms, optimizing operational workflows, or enhancing safety measures. These measures aim to prevent similar faults from recurring, thereby improving the robot's overall performance and user satisfaction.

In the event of a fault in a haircutting robot, a swift and effective response mechanism is equally important [18]. The robot system should include automatic fault detection capabilities and be able to immediately notify operators or technical support teams when a fault occurs. Additionally, the system should be designed with rapid fault troubleshooting and recovery procedures to minimize the impact of faults on service interruptions.

To achieve quick recovery, haircutting robots can adopt modular designs, allowing key components to be replaced swiftly. Furthermore, maintenance teams should conduct fault simulation training to enhance their ability to respond to various emergency scenarios.

Finally, fault analysis and improvement for haircutting robots should be an ongoing process. As technology evolves and user needs change, the robot system should continuously adjust and optimize its performance. By establishing a continuous improvement mechanism, such as regularly reviewing and updating maintenance protocols, continuously collecting user feedback, and leveraging the latest technological research, haircutting robots can ensure sustained high-efficiency, safety, and reliability in their operations over the long term [19].

References

1. Goodfellow, I., Bengio, Y., & Courville, A. (2016). *Deep learning*. MIT Press.
2. Levine, S., Pastor, P., Krizhevsky, A., & Quillen, D. (2018). Learning hand-eye coordination for robotic grasping with deep learning and large-scale data collection. *The International Journal of Robotics Research, 37*(4–5), 421–436.
3. Zhang, Y., & Yang, Q. (2020). A survey on multi-task learning. *IEEE Transactions on Knowledge and Data Engineering, 34*(12), 5586–5609.
4. Graves, A., Mohamed, A.-R., & Hinton, G. (2013). Speech recognition with deep recurrent neural networks. In *IEEE International Conference on Acoustics, Speech and Signal Processing* (pp. 6645–6649).
5. He, X., Liao, L., Zhang, H., Nie, L., Hu, X., & Chua, T.-S. (2017). Neural collaborative filtering. In *Proceedings of the 26th International Conference on World Wide Web* (pp. 173–182).
6. Krizhevsky, A., Sutskever, I., & Hinton, G. E. (2012). ImageNet classification with deep convolutional neural networks. *Advances in Neural Information Processing Systems, 25*, 1097–1105.
7. Shotton, J., Sharp, T., Kipman, A., Fitzgibbon, A., Finocchio, M., Blake, A., Cook, M., & Moore, R. (2013). Real-time human pose recognition in parts from single depth images. *Communications of the ACM, 56*(1), 116–124.
8. Mikolov, T., Karafiát, M., Burget, L., Cernocký, J., & Khudanpur, S. (2010). Recurrent neural network based language model. *Interspeech, 2010*, 1045–1048.
9. Mehdipour Ghazi, M., & Kemal Ekenel, H. (2016). A comprehensive analysis of deep learning based representation for face recognition. In *Proceedings of the IEEE Conference on Computer Vision and Pattern Recognition Workshops* (pp. 34–41).
10. Rybski, P., Anderson-Sprecher, P., Huber, D., Niessl, C., & Simmons, R. (2012). Sensor fusion for human safety in industrial workcells. In *2012 IEEE/RSJ International Conference on Intelligent Robots and Systems* (pp. 3612–3619). IEEE.
11. Sidobre, D., Broquere, X., Mainprice, J., Burattini, E., Finzi, A., Rossi, S., & Staffa, M. (2012). Human–robot interaction. *Advanced Bimanual Manipulation: Results from the DEXMART Project*, 123–172.
12. He, K., Zhang, X., Ren, S., & Sun, J. (2016). Deep residual learning for image recognition. In *Proceedings of the IEEE Conference on Computer Vision and Pattern Recognition* (pp. 770–778).
13. Zhang, Z., Miao, Y., & Wang, X. (2017). A comprehensive survey on face recognition based on deep learning. *Computer Vision and Pattern Recognition*, 345–358.
14. Ramesh, A., Pavlov, M., Goh, G., Gray, S., Voss, C., Radford, A., & Sutskever, I. (2021). Zero-shot text-to-image generation. In *Proceedings of the 38th International Conference on Machine Learning* (pp. 8821–8831).
15. Kim, J., & Park, M. (2019). Real-time safety monitoring in robotic systems using sensor fusion. *IEEE Transactions on Industrial Informatics, 15*(4), 2173–2181.
16. Calinon, S., Sardellitti, I., & Caldwell, D. G. (2010). Learning-based control strategy for safe human-robot interaction exploiting task and robot redundancies. In *IEEE/RSJ International Conference on Intelligent Robots and Systems* (pp. 249–254). IEEE.

17. Khalastchi, E., & Kalech, M. (2018). On fault detection and diagnosis in robotic systems. *ACM Computing Surveys (CSUR), 51*(1), 1–24.
18. Li, Z., Wang, Y., & Wang, K. S. (2017). Intelligent predictive maintenance for fault diagnosis and prognosis in machine centers: Industry 4.0 scenario. *Advances in Manufacturing, 5*, 377–387.
19. Achouch, M., Dimitrova, M., Ziane, K., Sattarpanah Karganroudi, S., Dhouib, R., Ibrahim, H., & Adda, M. (2022). On predictive maintenance in industry 4.0: Overview, models, and challenges. *Applied Sciences, 12*(16), 8081.

Testing, Validation, and Optimization 14

1 Testing Methods and Procedures for Prototypes

In the development process of haircutting robots, testing prototypes is a crucial step to ensure the system's functionality is complete and performance is stable. Through a series of testing methods and procedures, potential issues in system design and operation can be identified and resolved, laying the foundation for the optimization of the final product. The following sections will discuss in detail the testing of prototypes in aspects such as hardware, software, safety, and performance (Fig. 1).

1.1 Hardware Testing

The hardware system of a haircutting robot includes multiple components such as the manipulator arm, cutting tools, sensors, and cameras. The purpose of hardware testing is to ensure that each component operates normally as expected. The first step is functionality testing, which primarily verifies the basic functions of each hardware component. For example, testing the movement range of the manipulator arm, the cutting precision of the tools, and the environmental sensing capabilities of the sensors. The core of functionality testing is to ensure that each hardware module can operate independently and perform its respective tasks.

In addition to functionality testing, reliability testing is also necessary. Haircutting robots frequently perform certain operations during daily use, so testing their performance under long-term and high-frequency usage is crucial. By simulating common manipulator arm movements and the operation of cutting tools during the haircutting process, the durability of the hardware can be evaluated. For instance, conducting long-term operation tests on the manipulator arm and cutting tools to observe whether wear or precision degradation occurs after hundreds of operations [1].

© The Author(s), under exclusive license to Springer Nature Switzerland AG 2025

S. Li, *Haircutting Robots*, https://doi.org/10.1007/978-3-031-84026-5_14

Fig. 1 Haircutting robots must undergo rigorous comprehensive testing, including both software and hardware tests, as well as safety assessments

1.2 Software Testing

Software testing primarily includes evaluating system response speed and processing capabilities, as well as boundary condition testing. The software control system of a haircutting robot needs to maintain smooth operation under different environments and load conditions, especially when handling complex hairstyle designs. The system must be able to quickly generate cutting paths and respond to user operations, making performance testing under high-load conditions key to assessing its capabilities. For example, testing whether the software can rapidly respond to cutting commands during complex layered hairstyles to ensure a smooth haircutting process [2].

Simultaneously, boundary condition and extreme scenario testing are required. These tests aim to simulate the robot's performance under unconventional operating conditions to check system stability and fault recovery capabilities. For instance, testing how the system handles network interruptions, power failures, or sensor anomalies to ensure it can safely halt operations and resume normal functioning once conditions are restored. Such tests help developers identify potential system vulnerabilities and address them promptly.

1.3 Safety Testing

Ensuring that haircutting robots do not pose any harm to users under any circumstances makes safety testing a critical part of development. Firstly, mechanical safety testing includes monitoring whether the contact force between the manipulator arm and the customer's head remains within a safe range. For example, using force sensors to test the pressure applied by the manipulator arm during haircutting ensures that excessive force does not damage the customer's scalp or hair [3].

Secondly, electrical safety testing focuses on the insulation performance of the power system and leakage protection measures. To prevent accidents caused by electrical faults, the system should simulate power fluctuations or grounding issues during testing to verify whether it can quickly cut off the power supply and protect both the equipment and the user [4]. Additionally, network security testing ensures that the haircutting robot remains secure against hacker attacks, protecting both the user's personal data and preventing malicious software from controlling the robot to perform dangerous operations.

1.4 Performance Testing

The purpose of performance testing is to evaluate the haircutting robot's cutting precision, speed, and overall operational efficiency. In real-world scenarios, hairstylists and customers not only demand smooth operations but also aesthetically pleasing and consistent hairstyles. Therefore, testing the robot's performance under different hairstyle designs, hair textures, and hair volumes helps developers optimize cutting path planning algorithms and improve cutting effects [5].

For example, performance testing can involve having the robot complete various types of hairstyle designs, from short to long hair and from straight to curly hair, to assess its performance in each scenario. Additionally, by measuring the time taken for haircutting and the range of errors, the robot's cutting efficiency and precision can be evaluated. The results of these tests directly influence the robot's final operational algorithms and hardware improvement directions.

1.5 Comprehensive Testing

After completing individual tests, the haircutting robot must undergo comprehensive testing. This step aims to evaluate the robot's overall performance in real-world usage scenarios. During comprehensive testing, the robot will face various complex environments and user demands, including diverse hairstyle designs, challenging environmental conditions (such as low lighting and high humidity), and prolonged continuous operation. Through this multi-dimensional testing, developers can thoroughly assess the system's stability, reliability, and user experience.

The ultimate goal of comprehensive testing is to ensure that the haircutting robot can handle various challenges in real-world operations and maintain high efficiency and stability over extended periods [6]. Comprehensive testing also allows for further adjustments and optimizations of the system to adapt to different user groups' needs.

2 User Trials and Feedback Collection

In the development and optimization process of haircutting robots, user trials and feedback collection are crucial stages. By involving hairstylists and ordinary users in actual usage, insights into the robot's performance in real scenarios can be gained from different perspectives. This feedback aids in optimizing and refining the system. User trials not only help developers identify potential issues but also provide data support for subsequent interaction design and user experience improvements.

2.1 User Experience Testing

User experience testing is the core task in the user trial phase. To comprehensively evaluate the practicality of haircutting robots, hairstylists, hairstyling designers, and ordinary consumers should be invited to perform actual operations and provide feedback on their operational experiences and system responses. During the testing process, hairstylists primarily use the robot to complete hairstyle designs and haircutting tasks, while ordinary users experience the role of customers, observing the haircut results and providing feedback.

Key aspects to focus on during user experience testing include:

1. Ease of Operation: Haircutting robots should feature simplified user interfaces and convenient operational workflows to reduce hairstylists' learning costs and usage difficulties. During testing, observe whether hairstylists can quickly master the robot's operational processes and whether additional training or technical support is required.

2. User Interface Friendliness: The user interface design must meet the requirements of intuitive operation. Does the interface provide clear feedback? Can hairstylists quickly adjust the cutting tools' settings through the interface? These questions are particularly critical during testing [7].

After user experience testing, gather feedback from different groups, especially hairstylists' opinions on the operational workflow and customers' evaluations of the haircut results. Hairstylists may focus on cutting precision, operational smoothness, while customers may focus on the hairstyle's aesthetic appeal and comfort. Through this feedback, system design shortcomings can be identified and provide a basis for subsequent optimizations.

2.2 Interaction Design Improvement

Based on feedback data from user experience testing, analyze potential interaction issues encountered during actual use and propose corresponding improvement suggestions. Interaction design improvements mainly revolve around optimizing the user interface and simplifying operational workflows.

For example, hairstylists might find certain operational steps overly cumbersome or feel that the interface feedback is not timely enough. To address these issues, adjustments can be made to the interface design, such as adding more intuitive operational options like graphical cutting path displays or intelligent suggestion systems. Additionally, the system's response speed and feedback mechanisms might need further optimization to ensure hairstylists receive timely visual feedback when adjusting tools [8].

Moreover, customer feedback is also crucial. If customers are dissatisfied with the final haircut during the experience process, the system can improve hairstyle recommendation algorithms, cutting path planning, and force feedback mechanisms. By categorizing and analyzing feedback from different user groups, areas for improvement across various aspects of the system can be identified, further enhancing the user experience.

2.3 User Satisfaction Analysis

User satisfaction analysis involves assessing the overall usage experience of hairstylists and customers through quantitative and qualitative methods. Quantitative analysis can be conducted via surveys, asking users to rate their satisfaction with aspects such as ease of operation, interface design, cutting results, and interaction with the robot. Qualitative analysis can be performed through interviews or focus group discussions to gain deeper insights into users' feelings and suggestions during actual use (Fig. 2).

Fig. 2 Hairstylists and customers provide feedback through quantitative and qualitative methods to optimize the haircutting robot experience

In user satisfaction analysis, feedback from hairstylists and customers should be considered separately. Hairstylists' satisfaction is typically related to the smoothness of operations, cutting efficiency, and precision, while customers are more concerned with the hairstyle's aesthetic appeal, comfort, and whether they felt safe and relaxed during the haircutting process. Through in-depth analysis of this feedback, directions for system improvements can be identified [7].

For example, if hairstylists generally report that certain operational steps are not flexible enough, the system can optimize dynamic adjustment functions for cutting paths to enhance operational smoothness. On the other hand, if customers mention comfort issues, improvements can be made to the cutting tools' force feedback and adjustment mechanisms to ensure a gentler cutting process.

2.4 Big Data Analysis

With the collection of vast amounts of user trial data, system developers can utilize big data analysis tools to deeply mine and analyze feedback. By processing massive datasets, potential issues in system design can be comprehensively identified, and optimizations based on user behavior data can be performed.

The core of big data analysis is to identify common problems and personalized needs. For example, if certain operational steps frequently result in user errors or suboptimal cutting outcomes, it may indicate design flaws in those areas. By analyzing data from different user groups, factors such as age, gender, hair texture, and color affecting the usage experience can be identified, and default settings or haircut recommendation algorithms can be adjusted accordingly to generate more accurate recommendation lists.

Furthermore, big data analysis can help developers predict future user needs. By analyzing user usage data and feedback, the system can anticipate potential hairstyle design trends and optimize the hairstyle recommendation system's design in advance. With big data support, the system's interaction design and cutting path planning can become more intelligent and personalized, enhancing user satisfaction (Fig. 3).

3 System Improvement and Performance Optimization

During the development of haircutting robots, system improvement and performance optimization are key steps to ensure that the device achieves optimal results in actual usage. By continuously adjusting hardware and software, and incorporating user feedback and testing data, the robot's performance in various scenarios can be enhanced, ensuring system safety and stability.

3.1 Hardware Optimization

Hardware optimization focuses on improving the design of the manipulator arm and cutting tools, as well as enhancing sensors. Based on testing data and user feedback, developers can identify shortcomings in the manipulator arm's operational precision and flexibility, and optimize the mechanical structure accordingly. For example, adjusting the manipulator arm's joint design to enhance its flexibility, allowing for smoother movements during haircutting to prevent unnecessary pressure on the customer's scalp [9].

Additionally, the design of cutting tools needs to be improved based on actual operation data. For example, developers can optimize the sharpness of the cutting tool blades and their operating speed based on feedback from different hair textures and volumes, enabling more efficient handling of complex hairstyles. Furthermore, sensor layouts and types should be adjusted based on environmental sensing needs. For instance, adding more force feedback sensors to ensure that the

Fig. 3 Big data analysis analyzes general trends and preferences based on vast customer experiences with haircutting robots

pressure applied to the customer's scalp remains within a safe range, or optimizing camera positions to obtain more accurate hairstyle information and user status [1].

3.2 Software Optimization

On the software front, the main areas of optimization include algorithm improvements and enhancing safety features. By optimizing cutting path planning algorithms, the system can improve response speed and operational precision when handling complex hairstyle designs. For example, integrating machine learning algorithms allows the system to more intelligently predict cutting paths and continuously adjust cutting trajectories based on operational errors to ensure smooth operations and high-quality haircutting [10].

Simultaneously, safety enhancements are also a focus of software optimization. In terms of network security, haircutting robots need to bolster their protective measures to prevent system breaches or data leaks. By incorporating encryption algorithms, enhancing authentication mechanisms, and improving fault recovery systems, the robot can quickly recover from malicious attacks or system failures, ensuring data security and system continuity [3].

3.3 Cutting Effect Optimization

The cutting effect directly relates to user satisfaction, thus optimizing cutting path planning and force feedback mechanisms is essential. Through user feedback and testing data, developers can understand which hairstyle designs and cutting paths may result in subpar cutting quality. For example, certain complex hairstyles may require more precise path planning algorithms and more stable force feedback systems to handle [4].

To enhance cutting effects, the system can optimize the planning of cutting paths to ensure precision and consistency in each step of the cutting process. At the same time, the force feedback mechanism should be adjusted based on the customer's scalp sensitivity to ensure that the robot does not cause discomfort during haircutting. Through these optimizations, the system can maintain high-efficiency cutting while providing more refined operations, thereby improving cutting quality [11].

3.4 Iterative Improvement

System optimization is not a one-time process but is achieved through repeated cycles of testing, feedback, and optimization. In each iteration cycle, developers first identify problems and shortcomings in the system based on user feedback and testing data, then perform targeted optimization adjustments. After optimization, the robot undergoes user trials and testing again to verify the effectiveness of the improvements and identify new issues. Through this iterative process, the performance of all aspects of the system gradually improves, achieving continuous optimization and refinement [12].

This iterative improvement applies not only to hardware and software optimizations but also to cutting algorithms, interaction design, sensor layouts, and other aspects. Through this continuous optimization process, haircutting robots can consistently enhance their operational efficiency and user satisfaction in real-world applications.

3.5 Predictive Maintenance

Integrating predictive maintenance functionalities is key to ensuring the haircutting robot remains stable and reliable during long-term use. By embedding sensors and data analysis tools into the system, developers can proactively identify potential failure risks. For example, by monitoring the vibration frequency of the manipulator arm or the wear condition of the cutting tools, the system can predict which components may fail in the future and perform maintenance or replacements in advance [13].

Predictive maintenance not only extends the system's lifespan but also reduces the risk of service interruptions caused by equipment failures. Through regular fault diagnosis and maintenance, the system can maintain optimal operating conditions, minimizing downtime and enhancing the continuity and reliability of the user experience [14].

References

1. He, K., Zhang, X., Ren, S., & Sun, J. (2016). Deep residual learning for image recognition. In *Proceedings of the IEEE Conference on Computer Vision and Pattern Recognition* (pp. 770–778).
2. Levine, S., Pastor, P., Krizhevsky, A., & Quillen, D. (2018). Learning hand-eye coordination for robotic grasping with deep learning and large-scale data collection. *The International Journal of Robotics Research, 37*(4–5), 421–436.
3. Zhang, Y., & Yang, Q. (2020). A survey on multi-task learning. *IEEE Transactions on Knowledge and Data Engineering, 34*(12), 5586–5609.
4. Graves, A., Mohamed, A.-R., & Hinton, G. (2013). Speech recognition with deep recurrent neural networks. In *IEEE International Conference on Acoustics, Speech and Signal Processing* (pp. 6645–6649).
5. Krizhevsky, A., Sutskever, I., & Hinton, G. E. (2012). ImageNet classification with deep convolutional neural networks. *Advances in Neural Information Processing Systems, 25*, 1097–1105.
6. Shotton, J., Sharp, T., Kipman, A., Fitzgibbon, A., Finocchio, M., Blake, A., Cook, M., & Moore, R. (2013). Real-time human pose recognition in parts from single depth images. *Communications of the ACM, 56*(1), 116–124.
7. Lindblom, J., Alenljung, B., & Billing, E. (2020). Evaluating the user experience of human–robot interaction. *Human-Robot Interaction: Evaluation Methods and Their Standardization*, 231–256.
8. Zhang, S., & Sun, X. (2022). Designing social interactions for learning personalized knowledge in service robots. In *International Conference on Human-Computer Interaction* (pp. 656–671). Springer.
9. Lasota, P. A., Fong, T., & Shah, J. A. (2017). A survey of methods for safe human-robot interaction. *Foundations and Trends® in Robotics, 5*(4), 261–349.
10. Collins, G. R. (2020). Improving human–robot interactions in hospitality settings. *International Hospitality Review, 34*(1), 61–79.
11. Jiang, M., & Zhang, L. (2015). Big data analytics as a service for affective humanoid service robots. *Procedia Computer Science, 53*, 141–148.
12. Golombek, R., Wrede, S., Hanheide, M., & Heckmann, M. (2011). Online data-driven fault detection for robotic systems. In *2011 IEEE/RSJ International Conference on Intelligent Robots and Systems* (pp. 3011–3016). IEEE.

13. Borgi, T., Hidri, A., Neef, B., & Naceur, M. S. (2017). Data analytics for predictive maintenance of industrial robots. In *2017 International Conference on Advanced Systems and Electric Technologies (IC_ASET)* (pp. 412–417). IEEE.
14. Fernandes, M., Corchado, J. M., & Marreiros, G. (2022). Machine learning techniques applied to mechanical fault diagnosis and fault prognosis in the context of real industrial manufacturing use-cases: A systematic literature review. *Applied Intelligence, 52*(12), 14246–14280.

Industrialization Prospects and Business Models

15

1 Market Analysis and Positioning

As an emerging intelligent device, haircutting robots are gradually entering people's lives and work environments. With the continuous development of global automation technology, the market potential for haircutting robots is immense. By analyzing market size, target users, competitive environment, and user needs, businesses can formulate clear market positioning and development strategies, thereby enhancing product competitiveness and acceptance.

1.1 Overview of the Haircutting Robot Market

The global haircutting industry has a vast market size. According to market research data, the global beauty and personal care industry's market size reached approximately $532 billion in 2020 and is expected to reach $680 billion by 2026. Within this large market, haircutting and hairstyling services occupy a significant portion. Although automation technology has made remarkable progress in manufacturing, healthcare, and other fields, its application in the haircutting industry is still in its early stages. However, with the continuous maturation of artificial intelligence and robotics technology, the market for haircutting robots is gradually expanding (Fig. 1).

In recent years, global demand for automated devices has been steadily increasing, especially in the post-pandemic era, where consumers' demand for contactless services has significantly risen, driving the proliferation of smart devices. As a type of intelligent service robot, haircutting robots feature intelligence, efficiency, and safety, aligning with current market trends. It is anticipated that in the coming years, as robotic technology advances and costs decrease, haircutting robots will gain wider application in high-end salons, chain haircutting stores, and the home market.

© The Author(s), under exclusive license to Springer Nature Switzerland AG 2025

S. Li, *Haircutting Robots*, https://doi.org/10.1007/978-3-031-84026-5_15

Fig. 1 Different user groups (such as salons, home users, chain stores) have strong demands for haircutting robots

1.2 Target Market Selection

To ensure the successful market entry of haircutting robots, it is essential to clearly define their target user groups. Based on market research results, haircutting robots are suitable for the following main markets:

High-End Salons: High-end beauty salons typically emphasize technological innovation and the enhancement of customer experience. Haircutting robots in these venues can attract clientele seeking high-quality experiences by providing precise and efficient haircutting services. Additionally, the introduction of intelligent devices can enhance the salon's technological appeal and attractiveness, strengthening its market competitiveness [1].

Chain Haircutting Stores: Chain haircutting stores cater to large customer bases where operational efficiency is crucial. Haircutting robots can help these establishments improve service efficiency and reduce operational costs through automated haircutting processes. Furthermore, the personalized services offered by haircutting robots (such as intelligent hairstyle recommendations and automated cutting paths) can increase customer satisfaction and repeat business.

Home Market: With the proliferation of smart home devices, an increasing number of consumers desire convenient home services. Home haircutting robots can assist users in self-service haircuts or provide services for family members, particularly suitable for households with children or the elderly. This market has immense potential, especially in developed countries with high smart device penetration rates.

1.3 Competitive Environment Analysis

Understanding the current competitive environment is crucial before entering the market. Currently, there are some intelligent haircutting tools available, but most remain at the level of automated cutting devices and have not yet developed into complete haircutting robot systems. For example, some automated haircutting devices only feature automatic length adjustment and intelligent sensing functions. While they enhance efficiency to some extent, their functionality and intelligence levels are still not comparable to those of haircutting robots.

The main competitors for haircutting robots may come from the following areas:

Traditional Haircutting Equipment Companies: Many traditional haircutting tool manufacturers are transitioning towards intelligent devices, introducing equipment with automatic adjustment and intelligent recognition functions but have not fully entered the robotics field.

Start-up Technology Companies: Some start-ups focusing on artificial intelligence and robotics technology are developing intelligent haircutting devices, but most products are still in the prototype or early testing stages.

Through research and analysis of existing competitors' products, it can be observed that haircutting robots still possess significant differentiation advantages in the market. For instance, haircutting robots can integrate multiple intelligent functions such as cutting path planning, force feedback control, and personalized hairstyle recommendations to provide an integrated solution, which existing competitors find challenging to achieve [2].

1.4 User Needs Analysis

User needs analysis is the core aspect of product design and promotion. To ensure that haircutting robots can successfully enter the market, it is essential to deeply understand consumers' needs and pain points. Currently, consumers' acceptance of haircutting automation is gradually increasing, especially among high-end customer groups where technological appeal and personalized services are key factors attracting them.

Through user research, the following key needs have been identified:

Efficiency and Precision: Many consumers express a desire for more efficient haircutting processes while ensuring the precision and delicacy of hairstyles. Haircutting robots can meet users' dual requirements for efficiency and precision through intelligent cutting path planning and precise force feedback mechanisms.

Personalized Services: Consumers have diverse hairstyle design needs. Haircutting robots can leverage big data analysis and artificial intelligence algorithms to provide personalized hairstyle recommendations. They can even use virtual reality technology to allow users to preview hairstyle effects in advance, thereby increasing satisfaction.

Safety and Comfort: Some users, particularly household users and families with children, place high importance on the safety of the haircutting process. Haircutting robots can ensure operational safety through multi-level safety mechanisms (such as force feedback and emergency stop mechanisms), thereby enhancing user trust in the product.

Based on the above needs analysis, the positioning of haircutting robots in the future market should focus on providing efficient, personalized, and safe haircutting services. By employing precise user targeting and differentiated competitive strategies, successful market expansion can be achieved.

2 Business Operation Models and Profit Strategies

To ensure the successful promotion of haircutting robots in the market, the design of business operation models and profit strategies is crucial. By selecting suitable promotion models, formulating reasonable pricing strategies, and providing value-added services, businesses can better seize market opportunities and achieve profitability.

2.1 Direct Sales and Partnership Models

After entering the market, haircutting robots can be commercialized through various promotion models. The most common promotion models include direct sales, partnerships with chain stores, and B2B and B2C models.

Direct Sales Model: In this model, companies directly provide haircutting robot services to consumers by setting up their own stores. This allows businesses to have full control over the entire operational process, ensuring consistent service quality and brand image. This model is suitable for initial market promotion, allowing for rapid brand establishment and direct customer connections. However, the direct sales model incurs high operating costs, especially when conducting business in expensive prime commercial areas, requiring businesses to bear substantial fixed costs.

Partnership with Chain Stores: Partnering with existing chain haircutting stores is another effective promotion method. By establishing cooperative relationships with chain stores, companies can integrate haircutting robots into existing haircutting service systems, leveraging the market coverage and brand influence of chain stores to quickly penetrate the market. The advantage of this model lies in fast market penetration and lower initial costs, but businesses need to share profits with chain stores, which may affect profit margins.

B2B and B2C Models: In the B2B model, companies primarily sell or lease haircutting robots to other businesses (such as chain haircutting stores, hotels, fitness centers, etc.), while the B2C model involves directly providing products or services to end consumers. The advantage of the B2B model is larger transaction sizes and more stable target customer groups, whereas the B2C model allows direct access to end users, obtaining higher market feedback and brand loyalty [3].

2.2 Sales and Leasing Models

In the commercialization process, companies can choose between direct sales or leasing models. These two profit models each have their own advantages and disadvantages and need to be weighed based on market demand and product positioning.

Direct Sales Model: In this model, companies sell haircutting robots as products to end users (e.g., haircutting salons or individual consumers). This allows businesses to obtain higher one-time revenue and gives users full control over the robots. However, direct sales may limit the user base, especially for higher-priced haircutting robots, as many small and medium-sized salons or household users might find it unaffordable [4] (Fig. 2).

Leasing Model: The leasing model offers users lower initial costs, enabling more user groups to adopt haircutting robots. Companies can obtain stable cash flows through long-term leasing contracts and enhance customer loyalty by providing maintenance and services. The leasing model also allows businesses to collect user data for product improvements. While the leasing model brings ongoing revenue, it requires significant initial investment, as companies need to bear the costs of equipment maintenance and management [5].

2.3 Value-Added Services and Subsequent Profit

To increase profit margins and customer loyalty, companies can achieve subsequent profitability by providing value-added services. For example, companies can offer regular software upgrade services for haircutting robots to ensure that robots remain competitive in the ever-changing market demands. As artificial intelligence technology advances, companies can continuously optimize the hairstyle recommendation algorithms of haircutting robots and apply new functions and technologies to existing devices through software updates [6].

Additionally, companies can provide maintenance services and training courses to help hair salons or individual users better utilize the robots. Specifically for

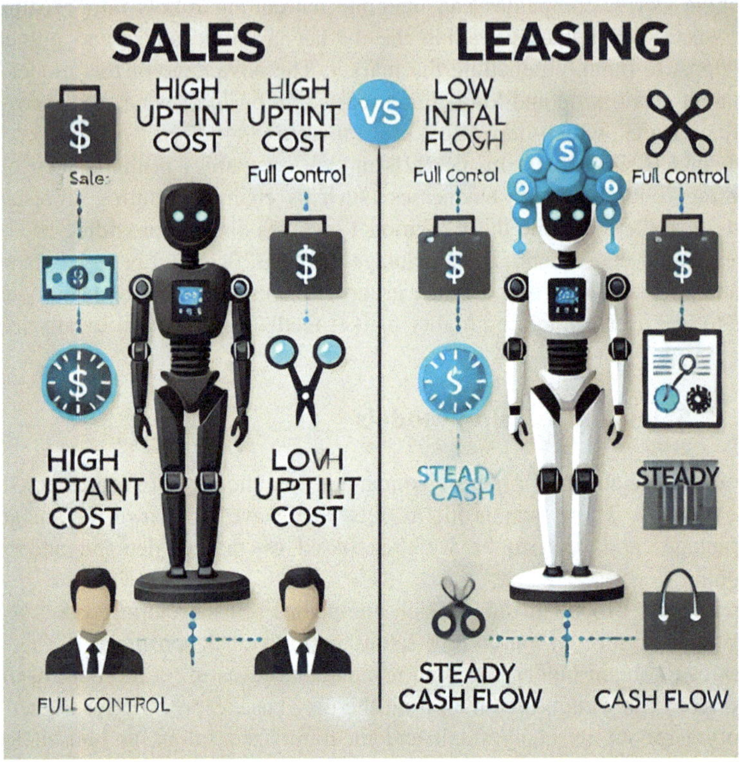

Fig. 2 Sales and leasing models for haircutting robots

hair salons, offering targeted technical support and training can help improve equipment usage efficiency and increase brand loyalty.

Another value-added service is on-demand services, such as AI-based personalized hairstyle recommendation systems. This service can automatically generate the most suitable hairstyle based on the user's facial features, hair texture, and personal preferences, and allow users to preview haircut effects through virtual reality technology in advance. These personalized services not only enhance user experience but also create additional revenue streams for companies [7].

2.4 Pricing Strategies and Market Promotion

The reasonableness of pricing strategies directly affects product market acceptance and profitability. Companies need to formulate pricing strategies based on market research and users' purchasing power to ensure that the product remains competitive in the market.

Cost-Plus Pricing: Companies can set product prices based on the actual production and operational costs of haircutting robots, combined with a reasonable profit margin. This pricing method is highly transparent and facilitates businesses in controlling costs and profits.

Value-Based Pricing: Another pricing strategy is based on the market value of the product, determining prices according to consumers' demand and perceived value of haircutting robots. For example, if consumers believe that haircutting robots can significantly improve haircut efficiency and quality, they are willing to pay higher prices [8].

In terms of market promotion, companies can adopt a combination of online and offline methods. Online promotion primarily involves social media, e-commerce platforms, and official brand websites, utilizing digital marketing techniques to increase brand exposure and user interaction. Offline promotion involves setting up experience stores, participating in industry exhibitions, and conducting training courses, allowing consumers to directly experience the convenience and technological appeal of haircutting robots [9].

3 Investment and Cooperation Opportunities

In the process of industrializing haircutting robots, investment needs and cooperation opportunities are critical factors for ensuring project success. Through the rational allocation of funds, the selection of financing models, and collaboration with strategic partners, companies can drive the rapid development of haircutting robots in the market. Additionally, global market expansion opportunities also bring more possibilities for industrialization.

3.1 Investment Needs Analysis

Industrializing haircutting robots requires substantial financial support, primarily concentrated in key areas such as technology research and development, production manufacturing, and market promotion.

Technology Research and Development: Technology research and development is the core of successfully entering the market with haircutting robots. To ensure product competitiveness, companies need to continuously invest in R&D to improve robot performance, optimize software algorithms, and enhance user experience. According to industry research, R&D costs for automated devices typically account for 40–60% of the total budget, covering areas such as artificial intelligence technology development, hardware design, and sensor optimization [10].

Production Manufacturing: After product development, companies need to invest in the production manufacturing phase. Haircutting robots involve multiple high-tech fields, including manipulator arms, sensors, and software control systems, resulting in higher production costs, especially during initial mass production when equipment procurement and manufacturing process improvements are required. This investment typically includes factory construction, production line setup, and the establishment of quality control systems.

Market Promotion: Successful market promotion is key to gaining user recognition for haircutting robots. Companies need to invest heavily in advertising, building sales channels, and user education. Particularly in a field with high technical barriers, companies also need to demonstrate the product's value through experience stores, exhibitions, and training courses.

3.2 Financing Models

To meet the diverse funding needs at different stages of industrialization, companies can choose from various financing models:

Venture Capital: Venture capital is a common financing method, especially suitable for technology-driven start-ups. By attracting venture capital, companies not only gain financial support but also leverage the resources and experience of investors to accelerate the commercialization of the project. Venture capital typically focuses on the early stages of technology R&D and market promotion, providing necessary startup funds.

Crowdfunding: Crowdfunding is a newly emerging financing method, particularly suitable for innovative hardware products. Through crowdfunding platforms, companies can showcase haircutting robots to the market in advance, attracting investment from potential users. The advantage of this model is that it not only provides financial support but also establishes brand recognition and collects user feedback at an early stage.

Corporate Partnerships: Corporate partnerships involve collaborating with other companies to obtain funds and resources. In haircutting robot projects, companies can partner with automation equipment suppliers or leading haircutting industry giants, reducing technology development and production costs through technological collaboration and shared manufacturing. Additionally, partnerships with chain haircutting enterprises can provide stable orders and customer resources, further enhancing financing capabilities.

3.3 Strategic Partners

Identifying suitable strategic partners is crucial for advancing the industrialization of haircutting robots. Through collaboration with strategic partners, companies can share resources, enhance technological capabilities, and accelerate product market entry.

Haircutting Industry Giants: Establishing partnerships with leading companies in the haircutting industry can help companies quickly penetrate the haircutting market. These companies possess mature market channels and extensive customer resources, allowing businesses to rapidly gain market share and reduce promotional costs through collaboration.

Automation Equipment Suppliers: Automation equipment suppliers can provide core technological support for haircutting robots, especially in the areas of manipulator arms and sensors. By collaborating with suppliers, companies can lower technology development and production costs while ensuring that product technology levels remain industry-leading [11].

AI Technology Companies: Artificial intelligence is a core technology for haircutting robots. By partnering with AI technology companies, businesses can continuously optimize their algorithms and enhance the intelligence levels of haircutting robots. Additionally, AI companies can help develop personalized hairstyle recommendation systems based on big data and deep learning, increasing the product's competitiveness [12].

3.4 Global Expansion Opportunities

Haircutting robots not only have broad application prospects in domestic markets but also possess ample opportunities in global markets. With the proliferation of robotics technology and smart homes, haircutting robots are expected to find market demand in various countries and regions worldwide (Fig. 3).

Developed Countries Market: In developed countries such as the United States, Europe, and Japan, the penetration rate of smart devices is high, and consumers have a strong acceptance of new technologies. Haircutting robots can first enter these markets, providing services to high-end salons, chain haircutting stores, and smart home markets. Additionally, consumers in these countries place greater emphasis on personalized services, which haircutting robots can fulfill through intelligent hairstyle recommendations and efficient haircutting experiences.

Emerging Markets: In emerging markets such as China, India, and Southeast Asia, the haircutting industry is still predominantly manual, but consumer demand for automated services is gradually increasing. These markets are becoming more receptive to haircutting robots. Companies can adopt flexible pricing strategies to enter emerging markets, capturing market opportunities early.

Localization Adaptation Strategies: When entering global markets, companies need to adapt to different cultural habits, haircutting needs, and regulatory requirements. For example, in terms of hairstyle preferences, language support, and usage habits, companies need to adjust their products accordingly to better meet the user needs of different regions.

Fig. 3 Haircutting robots are full of opportunities in the global market

References

1. Chiang, A. H., & Trimi, S. (2020). Impacts of service robots on service quality. *Service Business, 14*(3), 439–459.
2. Fosch-Villaronga, E. (2019). *Robots, healthcare, and the law: Regulating automation in personal care*. Routledge.
3. Paluch, S., Wirtz, J., Kunz, W. H. (2020). Service robots and the future of services. *Marketing Weiterdenken: Zukunftspfade für eine marktorientierte Unternehmensführung*, 423–435.
4. Elizondo-Noriega, A., Tiruvengadam, N., Güemes-Castorena, D., & Beruvides, M. G. (2022). A system dynamic-based archetype for capital leasing of industrial robots. In *2022 Portland International Conference on Management of Engineering and Technology (PICMET)* (pp. 1–20). IEEE.
5. Salo, M., Rantala, T., & Hakanen, T. (2023). Innovative service robotics business models-industrial customer perspective. In *ISPIM Conference Proceedings. The International Society for Professional Innovation Management (ISPIM)* (pp. 1–10).

6. Mandeep, S. (2012). *Building customer loyalty through value added services a case of telecom sector*.
7. Abdullah, Hussain, A., Ali, S., Kim, H. C., Sain, M., & Aich, S. (2021). Hybrid based model face shape classification using ensemble method for hairstyle recommender system. In *International Conference on Smart Computing and Cyber Security: Strategic Foresight, Security Challenges and Innovation* (pp. 61–68). Springer Nature Singapore.
8. Li, X., Huang, L., Dou, G., & Bian, Y. (2023). Service principal selection and pricing strategies: Service robots vs. human staff. *Human Staff*.
9. Paluch, S., Wirtz, J., & Kunz, W. H. (2020). Service robots and the future of services. *Marketing Weiterdenken: Zukunftspfade für eine marktorientierte Unternehmensführung*, 423–435.
10. Benmelech, E., & Zator, M. (2022). *Robots and firm investment*. National Bureau of Economic Research.
11. Huang, M. H., & Rust, R. T. (2021). Engaged to a robot? The role of AI in service. *Journal of Service Research, 24*(1), 30–41.
12. Wong, K. L. Y., Hung, L., Wong, J., Park, J., Alfares, H., Zhao, Y., Mousavinejad, A., Soni, A., & Zhao, H. (2024). Adoption of artificial intelligence–enabled robots in long-term care homes by health care providers: Scoping review. *JMIR Aging, 7*(1), e55257.

The manufacturer's authorised representative in the EU is Springer Nature Customer Service Centre GmbH, Europaplatz 3, 69115 Heidelberg, Germany. If you have any concerns regarding our products, please contact ProductSafety@springernature.com

Printed and bound by CPI Group (UK) Ltd, Croydon, CR0 4YY
30/09/2026
02225610-0005